Communications in Medical and Care Compunetics

Volume 1

Series Editor
Lodewijk Bos, International Council on Medical and Care Compunetics, Utrecht, The Netherlands

For further volumes:
http://www.springer.com/series/8754

ICMCC

The international council on medical & care compunetics

This series is a publication of the International Council on Medical and Care Compunetics.

International Council on Medical and Care Compunetics (ICMCC) is an international foundation operating as the knowledge centre for medical and care compunetics (COMPUting and Networking, its EThICs and Social/societal implications), making information on medicine and care available to patients using compunetics as well as distributing information on use of compunetics in medicine and care to patients and professionals.

Lodewijk Bos · Denis Carroll
Luis Kun · Andrew Marsh
Laura M. Roa
Editors

Future Visions on Biomedicine and Bioinformatics 1

A Liber Amicorum in Memory of Swamy Laxminarayan

Springer

Lodewijk Bos
International Council on Medical and Care Compunetics (ICMCC)
Stationsstraat 38
3511 EG Utrecht
The Netherlands
e-mail: lobos@icmcc.org

Denis Carroll
Head of KTP Unit
University of Westminster
Regent Street 309
London W1B 2UW
UK
e-mail: d.c.carroll@westminster.ac.uk

Luis Kun
Center for Information Assurance Education
National Defense University
300 5th Avenue SW, Marshall Hall
Washington, DC
20319-5066
USA
e-mail: kunl@ndu.edu

Andrew Marsh
VMW Solutions Ltd
Northlands Road, Whitenap 9
Romsey SO51 5RU
UK

Laura M. Roa
Departamento de Ingenería de Systemas y Automática
University of Seville
E.T.S. de Ingenieros Indus
Camino Descubrimientos s/n–Isla Cartuja
41092 Sevilla
Spain
e-mail: laura@esi.us.es

ISSN 2191-3811
e-ISSN 2191-382X
ISBN 978-3-642-15050-0
e-ISBN 978-3-642-15051-7
DOI 10.1007/978-3-642-15051-7
Springer Heidelberg Dordrecht London New York

Cover design: eStudio Calamar S.L.

Printed on acid-free paper

Springer is part of Springer Science+Business Media (www.springer.com)

Contents

A Tribute to Swamy

Lodewijk Bos

"Thank you so much for your note of invitation. I am very pleased to accept. I think this could be an exciting conference, it seems like a unique model. I look forward to working with you and Andy. I will certainly make it a point to get together with you, if i happen to be in Europe in the next couple of months."

This was the first email I received from Swamy Laxminarayan, 23 November 2003. Andy Marsh had linked us together. A month later he accepted to be scientific chair of the ICMCC Conference in 2004, the conference that led to the ICMCC foundation of which both were co-founders.

Within 2 months after that first message, he had invited quite a number of friends to support the conference. Many of those are still involved in some way with the goals and works of ICMCC.

I was a complete stranger to the area of health technology. I have told the story many times; I wanted to organize a conference looking from the ICT angle towards the fields of medicine and care. As Swamy pointed out, a unique concept. In the only 22 months I had the privilege to know him and even call him a friend he listened to my ideas and supported and massaged them.

He opened doors for me, got ICMCC its membership of the IFMBE. I met him only five times in person; two ICMCC conferences, once at an IFMBE event in Italy, once for a meeting in London and once, in 2004, when I visited him in Idaho. Looking back, it seems unbelievable that someone, in a long-distance friendship (the time difference between Utrecht and Idaho caused him to say that ICMCC never sleeps), could have such an impact. Due to his support and belief in what I wanted to achieve with ICMCC we still exist as a foundation, despite the fact that in the meantime two more members of our board passed away and I suffered from a very aggressive cancer. During my studies in the arts as a young man Iwas taught

L. Bos (✉)
Utrecht, The Netherlands
e-mail: lobos@icmcc.org

Commun Med Care Compunetics (2011) 1: 1–2
DOI: 10.1007/8754_2010_11

Published Online: 6 January 2011

by the best in the world, but none of these great teachers had such an influence as this modest friend from Idaho.

He knew how to bring people together and for our second (and his last) ICMCC Event in 2005 he managed to bring together the president-elect of the IEEE (Prof. Michael Lightner), the president of the IFMBE (Prof. Dr. Joachim Nagel) and the president of the IEEE-SSIT (Prof. Brian O'Connell). And he definitely enjoyed it.Michael Lightner, Joachim Nagel, Jeremy Nettle, Winnie Tang, Lodewijk Bos, Swamy Laxminarayan, Brian O'Connell (ICMCC Event 2005)

This book is a tribute and an archive. Part 1 forms the tribute, where, 5 years after his death, some of his friends and colleagues give an impression of their work to date.

Part 2 is the archive, as we re-publish some of the last papers to which Swamy contributed, trying to show the present reader the width of his horizon. And finally an overview of what Swamy has done during his live, including an almost complete, more than impressive bibliography.

In grateful memory,

Lodewijk Bos
President ICMCC

A Comprehensive View of the Technologies Involved in Pervasive Care

Laura María Roa Romero, Luis Javier Reina Tosina, Miguel Ángel Estudillo Valderrama, Jorge Calvillo Arbizu and Isabel Román Martínez

Abstract It is widely accepted that the application of Information and Communications Technologies (ICT) in the healthcare environment leads to an improvement in medicine and healthcare delivery. The ageing of population, the prevalence of chronic condition, and other societal changes, as well as advancements in science and technology, require an evolution in healthcare delivery from centralized, general and reactive care, towards distributed, personalized and preventive care. The application of ICT can address these new scenarios but it is needed a methodological approach to establish common guidelines so that the developed systems are interoperable, reusable and future-proof. In this chapter, we present a methodology based on Open Distributed Processing (ODP) and standards to address the complexity of design and development of distributed systems in healthcare. This methodology specifies systems according to decomposition in viewpoints, each one focused on particular issues. We test this method by applying it to the general healthcare domain and particularizing it for a specific use case of an ICT application, a pervasive care system. We describe both the technology neutral viewpoints and those dependent on it.

L. M. Roa Romero (✉), M. Á. Estudillo Valderrama and J. Calvillo Arbizu
CIBER de Bioingeniería, Biomateriales y Nanomedicina (CIBER-BBN),
Biomedical Engineering Group, University of Seville, Seville, Spain
e-mail: laura@esi.us.es

L. J. Reina Tosina
CIBER de Bioingeniería, Biomateriales y Nanomedicina (CIBER-BBN),
Department of Signal Theory and Communications, University of Seville, Seville, Spain

I. Román Martínez
CIBER de Bioingeniería, Biomateriales y Nanomedicina (CIBER-BBN),
Area of Telematic Engineering, University of Seville, Seville, Spain

Commun Med Care Compunetics (2011) 1: 3–19
DOI: 10.1007/8754_2010_8

Published Online: 16 November 2010

1 Introduction

Nowadays it is widely accepted that the application of Information and Communications Technologies (ICT) in the healthcare environment lead to the improvement of medicine and healthcare delivery. Thus, among the potential benefits, it can be remarked a more efficient management of health information; the possibility of remotely collecting data from wearable monitoring devices; more sophisticated and accurate techniques and methods for treatment and diagnosis; and so on.

In the near future, ICT will allow transferring to the clinical practice all the advancements in science and technology since the last decade (genomic and proteomic data, molecular images, miniaturization of monitoring devices, implanted sensor units, nanotechnologies...) in order to enhance the prevention, diagnosis, and treatment of diseases. In the meantime, a revolution in medical methods is being carried out towards a more personalized care. There are more and more resources centered on a single subject of care, for example, body sensor networks for real-time monitoring, studies or analysis made over different scales allowing a multilevel diagnostic, models and simulation tools for predicting drugs reactions or diseases evolution, etc. The application of ICT in healthcare will ease the current healthcare delivery evolve from using partial and isolated information to synthesizing all the knowledge available about each person in a cohesive whole.

Advancements and opportunities have led to the public awareness of the need for a new reorientation of health resources, creating a new conscience in the citizen toward healthcare delivery, who claims for the practice of a preventive medicine instead of the prevailing reactive medicine. Patients demand a pervasive care, more information and knowledge for a personalized healthcare, aimed to an improvement in the quality of life. Furthermore, citizens want to be more and more involved in their own healthcare and maintenance of well-being that will ease, for example, behavioral changes in their daily life (eating habits, physical activity routines...) to prevent or treat possible diseases.

Governments have not been unaware of the impact of ICT as a key factor to derive cost-effective solutions in this changing scenario, and have been forced to develop scientific policies to address the new challenges. Examples of these are the Ambient Assisted Living Joint Programme, launched by the European Commission, or the Spanish Law on the promotion of personal autonomy and care for dependent persons (Ley 39/2006, de 14 de diciembre, de Promoción de la Autonomía Personal y Atención a las personas en situación de dependencia; 2006 Dec 14).

This change of scenario is going to be sped up due to a set of additional growing concerns. During the last two decades different institutions and authorities have warned against a collapse of the public health system in the developed world, by the middle of the century, due to the confluence of diverse factors, including the population ageing, the prevalence of the chronic condition in extensive population groups, the change of social models and structure of family, the impact of migration and population movements in the outbreak of infectious diseases, and the corresponding growth in health expenditure. The application of ICT in healthcare in a methodological way will be a cornerstone to address the new scenarios and challenges, reducing the associated cost and enhancing the efficiency of assistance processes.

A clear example of application of ICT is the paradigm of telemedicine defined as the investigation, monitoring, and management of patients and the education of patients and staff using systems that allow ready access to expert advice and patient information, no matter where the patient or relevant information is located [1]. The concept of telemedicine has gradually been widened from the use of videoconferencing for remote consultation, toward collaborative medicine and research through distributing system capabilities over high-speed networks and information infrastructures.

The movement of health resources from a centralized scenario based on a hospital towards a distributed one across organizations boundaries, including also user home as a location where healthcare can be delivered, is acknowledged as a key issue giving responses to the growing healthcare needs [2]. One of the pioneers to envision this paradigm shift was the late Professor Swamy Laxminarayan [3], who predicted in 2002 the shift of the information age towards a knowledge-centric paradigm, through the opening of new frameworks for the integration of all the biophysical, biochemical, and physiological knowledge for prediction purposes. Hence, the telemedicine paradigm, which was first conceived as a feasible solution, in the sense of cost-efficiency ratio, to deploy health services at underserved areas, is progressively being shifted toward the concept of pervasive care systems (PCS). The centralized health model is thus moved to a distributed one, with the patient/citizen acquiring a more active role in the healthcare delivery process.

Under the direction of the cited editorial article, several research groups and projects [4–6] have made important advances in new settings for PCS. Although under the basis of tele-healthcare is the claim that health services can be offered more effectively and with less cost by providing connectivity with ICT [7] the utilization rates for homecare projects are still falling well below expectations. On the basis for such failure is the common practice to yield technological solutions to particular cases.

This problem affects not only to PCS but also to all healthcare scenarios where ICT have been applied. Due to design methodologies centered on particular requirements and use cases, tackling the specific problem isolated from the whole healthcare organization, consequent solutions lack of flexibility, scalability and reusability. They can only be used in the particular context for which they were designed as they have not taken into account other systems and the interoperability with them. This situation results in a healthcare environment with a wide spectrum of devices, systems and solutions; each one focused on one specific problem; implementing heterogeneous, often proprietary, technologies; with few or no possibilities of interoperability between them and reuse of their capabilities; and finally, with large funds invested in solutions that in a few years may become obsolete.

The seek for new methodological approaches is thus justified. Proper methodologies are necessary to design general architectures that can be suited to every particular case in the healthcare environment. The vision of this environment as a cohesive whole will allow future solutions and systems to cooperate between them in order to reuse capabilities and achieve more complex goals, easing a more efficient and personalized healthcare delivery integrating all the knowledge related to one single person from diagnosis and treatments to daily life, genomics and monitored data. Thus, each particular solution and system will be a building block of a bigger system (the healthcare environment) that is evolvable and more and more complex, sophisticated and efficient.

In this book chapter, we face the application of ICT within the healthcare environment from a comprehensive vision of the whole complexity and potential solutions and technologies to apply. We present a methodological approach based on the Open Distributed Processing Reference Model (ODP-RM) (ISO 10746-1, 2, 3, 4: Information technology—open distributed processing—reference model, 1996) to develop open architectures of health, well-being and social care services. We address the requirements of interoperability, reusability and scalability by conforming international standards and recommendations in the whole development process. This methodology is technology neutral and widely enough to cover all the current and future requirements of scenarios and domains within the healthcare environment from healthcare delivery in hospitals to well-being outdoors services or remote monitoring in homecare.

As a proof-of-concept, we apply the proposed methodology to the development and deployment of a pervasive care platform. We present a general architecture for homecare, developed by our group, which stems from the concept of a personalized care through knowledge generation and we particularize on the underlying technologies. By following the ODP methodology, we start with the analysis of particular requirements of homecare delivery in relation to the subjects of care. As the methodology is technology neutral, it can be applied to current and future technologies and, in particular, we focus on the different solutions with today's available technologies and short-term evolution, and paying special attention to the use of industry standards to benefit from interoperability between heterogeneous systems.

2 Materials and Methods

As it has been pointed out above, the healthcare environment is (and will continue) evolving from centralized scenarios to span across boundaries and domains becoming a complex distributed environment with several heterogeneous devices, systems and capabilities. Distributed systems can grant several benefits to healthcare delivery, but they present a set of issues to be addressed, mainly due to the complexity of their development and management.

The design, development, deployment, maintenance and evolution of a distributed system are highly complex tasks, involving an interdisciplinary, numerous and usually dynamic team. Consequently, the design could represent a substantial body of specifications needed to manage successfully the structure. The formalization of this structure is what we call system architecture. As a single engineering solution might not meet all requirements, this architecture must be flexible. Furthermore, since a single vendor may not have all of the answers, it is essential that the architecture, and any functions necessary to implement it, are defined through a set of standards, so that multiple vendors can collaborate in the provision of distributed systems. Such standards will enable to build open, integrated, flexible, modular, secure and easily manageable systems. Hence, taking into account all previous arguments, it is easy to understand that a coordinating framework for the standardization of the architecture is needed.

In a very simplified way we could say that such framework fundamentally consists of: a precise concepts language; a set of rules for the consistent structuring of system specifications; usually a set of fundamental or widely applicable functions for the construction of these systems; and several transparency prescriptions showing how to use these functions to hide users from the complexity of distribution. An efficient framework should allow different parts of the design to be developed separately if they are independent, but it should clearly identify those points where different aspects of the design constrain each other.

Different standards provide such framework for the formalization of distributed systems architectures. Two samples are the Reference Model of Open Distributed Processing (RM-ODP) and Model Driven Architecture (MDA) [8]. Although these standards differ in several aspects (like the use of viewpoints in ODP against a model approach in MDA), they have a similar philosophy; separating specifications targeting a technology-neutral viewpoint of the system, from those including details that specify how a particular underlying technology is used in the system. As a technology independent specification is suitable for a number of different platforms of similar type, this approach improves the interoperability between components designed by following the same specification although they were developed using different technologies.

One of the characteristics that makes more suitable RM-ODP over other standards is that it provides a coordinating framework for the standardization of open distributed processing that supports distribution, interoperation, platform and technology independence, and portability, together with an enterprise architecture framework for the specification of ODP systems. The framework for system specification provided by the RM-ODP has four fundamental elements: an object modeling approach for system specification; the specification of a system in terms of separate but interrelated viewpoint specifications; the definition of a system infrastructure providing distribution transparencies for system applications; and a framework for assessing system conformance.

By following principles as encapsulation and abstraction, RM-ODP allows the description of system functionality to be separated from details of system implementation, allowing the hiding of heterogeneity, the location of failure, the implementation of security and the hiding of the mechanisms of service provision from the service user. Moreover, the basic characteristics of heterogeneity and evolution imply that different parts of a distributed system can be purchased separately, from different vendors.

The concept of RM-ODP viewpoints framework, therefore, is to provide separate viewpoints into the specification of a given complex system. Each of these viewpoints satisfies an audience with interest in a particular set of aspects of the system. Associated with a single viewpoint is a viewpoint language that optimizes the vocabulary and presentation for the audience of that viewpoint. RM-ODP defines five viewpoints: enterprise, information, computational, engineering, and technology. A viewpoint (on a system) is an abstraction that yields a specification of the whole system related to a particular set of concerns. The five viewpoints defined by RM-ODP have been chosen to be both simple and complete, covering all the domains of architectural design.

The RM-ODP is a general-purpose framework to design and develop standards architectures in any domain. Within the healthcare environment, the ISO/CEN 12967 standard (Health Informatics Service Architecture, HISA, ISO 12967-1, 2, 3, 2009) describes an architecture for the integration of healthcare information services whose base is the ODP methodology. It is pointed out that HISA does not aim to represent a final complete set of specifications. On the contrary, it only formalizes fundamental features that are common and currently essential in any advanced healthcare system, as well as relevant for any healthcare sector and usable by any application also for facilitating the mutual interworking. Therefore, the HISA is an open framework that can be extended during time according to the evolution of the healthcare organization. The specifications are formalized avoiding any dependency on specific technological products and/or solutions.

By using the RM-ODP and fulfilling the HISA requirements, open distributed systems can be developed to support the procedures of healthcare organizations assuring that different vendors will be able to provide interoperable hardware and software components, and even the interoperability between components from different organizations, whenever they follow this standard. As a result social, health care and well-being services for the comprehensive assistance to citizens could be provided by combining capabilities from different vendors and organizations.

3 Results

As it has been shown above, one of the keys of success of open and interoperable solutions is the adoption of standards and recommendations in both general and healthcare domains. Due to this, our methodological approach tries to be conformed to all the relevant standardization efforts in the implied issues.

We separate the results in two differentiated points. The first one is focused on the methodology covering the whole healthcare environment and how it can be adapted to different scenarios in order to develop interoperable systems and solutions, which are reusable and with a long-term life based on technology neutrality. The second point presents a proof-of-concept by applying the methodology to a specific scenario, a pervasive care system. This solution benefits from all the advantages of distributed systems from the very initial design step and it results in a fully interoperable system with other current and future developments, scalable, open and reusable.

3.1 Description of a Particular Use Case of ICT Application

In this section, a description of a Pervasive Care System (PCS) is shown by following the methodology stated. The classical architecture for PCS spanned around three geographically located scenarios—the home, the hospital and the information, data and service provider center—has progressively moved to a distributed network, in which the information sources are heterogeneous in nature, and the frontiers among the scenarios are diffuse. This poses the challenge of moving to the concept of domain, which can span different scenarios depending on the user context. Three domains (involving three or more scenarios) are common in all cases: the point of care (PoC) domain and the professional user (PU) domain; and both supported by an infrastructure domain granting distribution transparency and information exchange capabilities. A viewgraph showing the relationships within such domains is depicted in Fig. 1.

In the PoC domain the subject of care and his/her carers (either professionals or not) interact with devices and systems. Thus, it is required a seamless integration of technologies supporting both in-door and out-door operation, and distributed

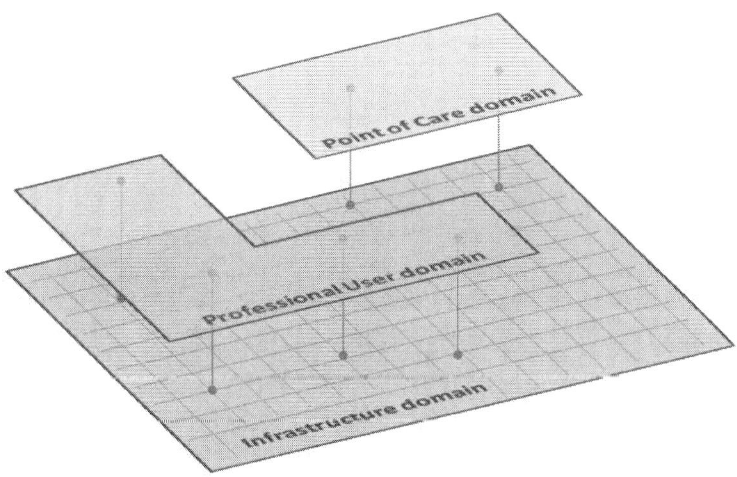

Fig. 1 Relation between domains of a Pervasive Care System

processing. In the PU domain, a global view of the health state of the subject of care must be provided to the professional user. This domain will be spanned across boundaries connecting heterogeneous systems and services from healthcare organizations and third-parties, e.g., EHR, PHR, clinical support systems, models and simulation tools, etc. All of them will be information sinks by using data generated in the PoC domain and by generating knowledge about the health state of the subject of care. Both domains will be supported by an ICT infrastructure domain, allowing the real-time processing of the data acquired from the subject of care, the efficient supervision and control of alarms, etc. In addition, security, confidentiality, privacy and protection for all data managed must be guaranteed.

In the next sections we apply the ODP methodology to the healthcare domain and this particular scenario, and we describe the ODP viewpoints.

3.2 Technology-Neutral Viewpoints

Due to the growing need of a complete architecture supporting the heterogeneity and complexity of healthcare services, our methodology is based on the RM-ODP and its particularization for the healthcare environment, the HISA standard. We have chosen ODP because of its maturity and complete methodology of distributed system specification. Moreover, it has been used to define HISA, a standard architecture applied in the healthcare environment, by following the same goals which we try to achieve.

As it has been described above, the key concept of the ODP framework is the viewpoint: an abstraction that yields a specification of the whole system related to a particular set of concerns. The five viewpoints (enterprise, information, computational, engineering and technology) cover all the domains of architectural design and they are not completely independent; key items in each one are identified as related to items in other viewpoints, allowing achieving a complete and cohesive specification of the whole healthcare environment.

Our methodology describes the healthcare environment from a general vision including all potential services, systems and scenarios related to well-being, health, and social care of end-users. Due to the complexity and heterogeneity of such domain, in this book chapter we only present an overview of the possibilities of RM-ODP as a formalization tool allowing the establishment of common guidelines and directives to a wide spectrum of services and systems.

In this section we analyze the three upper viewpoints of the healthcare environment—enterprise, information and computational—and we particularize them for the PCS. The viewpoints are going to be described by paying special attention to the internationally accepted standards suitable to be applied in health domain and by considering the casualty of current and future services. Thus, we propose a general-purpose technology-neutral future-proof methodology to develop and deploy open, scalable and interoperable systems in the healthcare domain. Because of the technological neutrality of the methodology, the two lower viewpoints (engineering and technology) are not tackled in this section, but they do in the following one where a specific PCS scenario is described.

It is unfeasible to make an exhaustive specification of the healthcare environment due to its complexity and heterogeneity between countries, organizations and even application scenarios. With this viewpoint (and the following two) we aim to state a set of directives to apply ICT in health domain in a proper and consistent manner.

In order to design and develop systems oriented to particular use cases and scenarios, the following results can be used and extended to address specific requirements. This process is tested and illustrated with the proof of concept of the PCS.

In an enterprise specification, an ODP system and the environment in which it operates are represented as a community. Thus, in our approach the complete healthcare domain is a community (which will include several sub-communities) and whose main objective is the prevention and treatment of citizens' illnesses and well-being by means of social and health care services; investigation and control of health public risks; and other methods and tools.

The processes performed in this community to achieve its objective are carried by entities fulfilling roles within the community. A rich set of potential entities and roles in health domain are presented in Table 1. The classification of roles in structural and functional ones conforms to ISO/TS 21298 (Health

Table 1 List of entities and corresponding roles in the healthcare environment

Entities			Roles
		Examples	
People	Professional	Any person	Structural (e.g., medical doctor, pharmacist, child-care worker…) Functional (e.g., responsible healthcare professional, administrator…)
	Non-professional	Any person	Functional (e.g., subject of Care, carer, guardian, parent…)
Organizations	Healthcare	Hospital Pharmacy Clinic	Healthcare deliver
	Others	Gym Dietary center Insurance company	Information source
Systems	Devices	Fall detection monitor Pulsometer	Information sink
	Data repositories	EHR PHR	Knowledge generator
	Other resources	Decision support system Models and simulation tools	

Table 2 Sample categorization of scenarios within the healthcare environment

Scenario categories	
Home	Indoor
	Outdoor
Healthcare	Hospital
	Primary healthcare center
	Peripheral therapy center
	Day hospital
	Nursing home
	…
Others	Ambulance
	Pharmacy
	Service provider third-party
	…

informatics—functional and structural roles, 2008). A sample of the wide spectrum of relevant scenarios within the healthcare domain is shown in Table 2. The correspondences between roles and scenarios are numerous and they are omitted for sake of concision.

Our methodology allows particularize the general specification to use cases. Thus, besides the general requirements of the health domain, it is needed to specify the particular goals, roles and requirements of (in this proof of concept) the PCS.

On the one hand, entities within this PCS fulfill one of the following roles. In the PoC domain two roles would be possible: the subject of care and the carer; the latter being a role usually associated to non-professional people. Notice that it should be possible to have more than one end-user playing the role of subject of care (e.g., in case of nursing homes instead of a particular home), what demands a scalable solution. Also it is possible have several carers. In the PU domain, the person in charge of the subject of care would be a general practitioner in a primary healthcare center, who would derive the case to a doctor/consultant in the hospital if required. Other scenarios can be included in this PCS such as the ambulance scenario under an alarm event (involving the subject of care and the hospital emergency unit).

On the other hand, several particular requirements can be added to those specified in the enterprise viewpoint of the general healthcare domain. First, the infrastructure domain must supply with several capabilities allowing the distribution of this system. Among them we should remark the access, failure, location, migration, relocation, and replication transparencies. They make it possible to implement ODP systems, which are distribution transparent from the point of view of users of those systems. Furthermore, the PCS requires that the infrastructure domain supports open standards in order to accomplish actual interoperability among heterogeneous systems and devices.

In the PU domain, a cohesive view of the health state of the patient must be provided by integrating all information and knowledge related to each subject of care. Thus, a heterogeneous spectrum of devices and systems across administrative and geographic boundaries must be interconnected and interoperable. These requirements are general of the healthcare distributed domain but especially relevant in the PU domain.

From the PoC domain perspective, it is required a seamless integration of technologies supporting both in-door and out-door operation, and distributed processing. Our approach to this requirement is based on the concept of intelligent multi-device platforms and multi-user platforms in the PoC. By following the current trends [9–11], the monitorization devices shall be unobtrusive, easy to use, hardly requiring manual intervention. The use of low-power wearable intelligent devices and sensors is thus completely justified. The power consumption minimization leads to portability of the system [12]. Other design guidelines for the intelligent biosensors include minimal obtrusion, implying restrictions on weight and size, calculation capabilities in order to achieve an effective distribution of the biosignals processing together with other devices within the PoC domain, and self calibration. On the other hand, distributing the processing capabilities across different layers by the multi-device and multi-user platforms eases the management of events and the discrimination/prioritization of alarms, which is another key objective.

Figure 2 presents a specific use case of the presented PCS. In particular, it describes a nursing home scenario with several subjects of care and carers, devices and remote connections to healthcare organizations and professionals, and diffused boundaries between PoC and PU domains.

As the second step of the methodology, we face the information viewpoint. The individual components of a distributed system must share a common understanding of the information they communicate when they interact, or the system

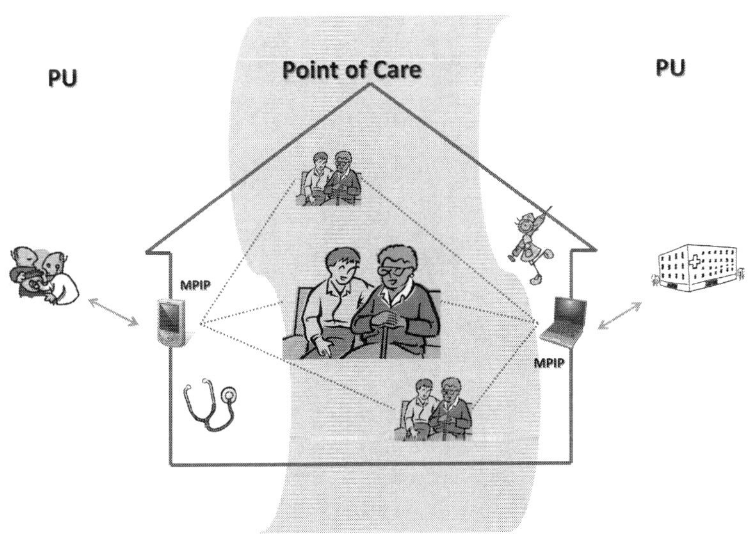

Fig. 2 Nursing home: same scenario shared by PoC and PU domains

will not behave as expected. The information viewpoint describes concepts for the specification of the meaning of information stored within, and manipulated by, an ODP system, independently of the way the information processing functions themselves are to be implemented.

In this viewpoint it has to be taken into account terminologies and concepts related to the healthcare domain, covering all the pieces of information that can be exchanged between systems. Due to the enormous complexity of the environment and the variety of scenarios, it is unfeasible to describe all the elements to consider. As a sample, the European standard EN13940 [13] defines a system of concepts to support the continuity of assistance, and as concrete existing terminologies to consider are, for example, those of clinical purpose (SNOMED [14], GALEN [15] ...) and those focused on testing and results (LOINC [16]).

By following our methodology, any terminology or information schema to include must be conformed to ODP methods and HISA schemas. As it has been stated above, HISA only formalizes fundamental features which are common and currently essential in any advanced healthcare system, thus it can be extended with any type of terminology and information schemas.

Finally, the computational viewpoint is directly concerned with the distribution of processing but not with the interaction mechanisms that enable distribution to occur (this will be specified in the technology viewpoint). The computational specification decomposes the system into objects performing individual functions and interacting at well defined interfaces.

To the general health domain, this viewpoint can be populated with the ODP functions (ISO 10746-1, 2, 3, 4: Information technology—open distributed processing—reference model, 1996) and the HISA computational specification. A more detailed viewpoint can be built when particular services and systems applied to specific scenarios are defined.

3.3 Technological Viewpoints

In this section, we present an analysis and brief design of the three domains described above. This part covers the main issues of ODP engineering and technology viewpoints.

The infrastructure domain must make it possible to implement ODP systems that are distribution transparent from the point of view of users of those systems, by using open standards and recommendations. There are several suitable underlying technologies such as Web services [17], Grid [18], and Cloud computing [19]. The broad acceptance of web services technology makes it an interesting tool but it lacks of complex capabilities. In other extreme, Cloud computing technologies provide a sophisticated set of capabilities but it is a new born technology and it lacks of maturity. In the middle, we can find Grid technologies, enhancing Web services features and being a proper infrastructure to support our PCS. Besides, in order to extend the coverage range of services to out-door scenarios, it will be interesting to establish gateways to portable communications platforms.

In addition to the characteristics and capabilities inherent to Grid technologies, the infrastructure must be conformed to a wide spectrum of standards to address security and confidentiality issues in order to protect such a sensible in nature information. An example of relevant standards in this issue are: ISO/IEC 10181 (describing the organization of security frameworks and defining required common security concepts) (ISO/IEC 10181-1, 2: Information technology—open systems interconnection—security frameworks for open systems, 1996), the ITU-T X.509 recommendation (defining a public-key and attribute certificate framework, (ITU-T X.509: The directory: public-key and attribute certificate frameworks, 2008), and ISO/TS 22600 (supporting the needs of healthcare information sharing across unaffiliated providers, organizations, subjects of care, and so on) (ISO/TS 22600-1, 2: Health informatics—privilege management and access control, 2006). A privilege management infrastructure based on standards to healthcare domain was analyzed in a previous work [20].

In the PU domain it is needed to ease the work of the professional user, often non expert in technology issues. Although the WWW has become the unified and easy-to-use interface to these advanced communication services, the design must be alert to the latest advances in man–machine interfaces [21]. Moreover, given the volume of data at the disposal of the professional user and the increasing number of members (either professional or not) belonging to this domain associated to a single subject of care, it is necessary to advance in the research and development of clinical decision support systems, and the generation of personalized clinical guidelines [22, 23]. The extended use of multiscale computational models will be essential to provide a better understanding of pathologies and test the efficacy of treatment guidelines before being directly applied to the patient.

Finally, the PoC domain defends the use of standards through the proposal of intelligent platforms for health monitoring. Figure 3 shows the proposed architecture of the PoC domain consisting in: a set of intelligent sensors (wearable and not) embedded in the domain, the intelligent platforms—multidevice intelligent platforms (MDIP) and multi-person intelligent platform (MPIP)—and their connections with the infrastructure domain. Despite two different platforms are defined, they do not have to correspond with different physical devices ought to their modular design that eases its technological implementation in several portable or fixed devices or just a single one, like a personal computer, a PDA, a regular mobile phone or even an ad-hoc preindustrial prototype.

The MDIP is conceived as a wireless hub that connects to different sensors forming a Wireless Sensor Network (WSN). Besides, the MDIP establishes communications with a gateway (MPIP) that links the PoC domain with the infrastructure domain. In both cases it is obvious the need to count on a suitable last-mile technology linking the home with other scenarios such as wide area network (WAN) wireless and wired technologies for short distance and short distance communications, respectively. The MDIP is mainly focused in the

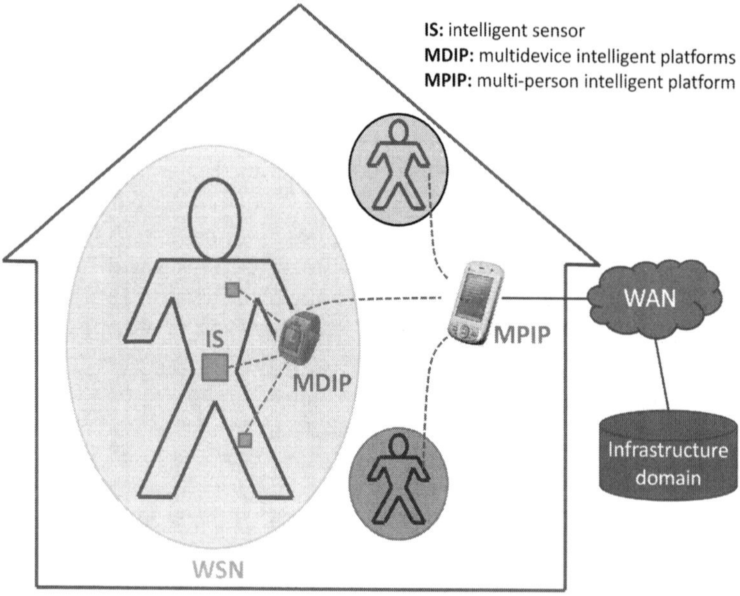

Fig. 3 Distributed architecture in the Point of Care domain

control and management of the intelligent biosensors attached to the subject of care while the MPIP develops a real-time multiuser monitoring task. For this purpose, the main functions of the latter taking advantage on the multilayer distributed paradigm in the PoC domain are: end-user identification and multi-person data flow management in real-time, services definition and implementation, management and prioritization of alarms, management and intelligent processing of the information, technology independence in any context and scenario, and biomedical information standardization. A deeper analysis of this platform is being object of a patent of invention together with a paper to be soon published by our Group.

The multimodal functionality of the MDIP refers to its ability to support different sensors, which will be selected according to the monitorization requirements of the end-user. The personalization of the device attends to its intelligence to adapt the event detection in the PoC domain to the user and the context of application for the sake of a better assistance of the end-user. This is developed by means of customized processing units in the MDIP executed in real time, which adapt their thresholds and internal variables transparently to the end-user through optimization algorithms constantly fed by the captured biosignals. Other features to be provided by the MDIP for the optimal operation of the system are: self-management and device–network control, management of energy aware strategies, sensor signal processing and management, communications management with the sensors (downlink) and the MPIP (uplink) and easy user interfacing. These features were taken into account in an implementation of the MDIP for a particular case of use: the fall detection [24].

4 Conclusions

Current advancements in science and medicine have led to the public awareness of the need for a new reorientation of health resources, creating a new conscience in the citizen who claims for the practice of a preventive medicine, more personalized and in which they can play an active role.

It is widely accepted that the application of Information and Communications Technologies (ICT) in the healthcare environment lead to the improvement of medicine and healthcare delivery, and this can be the engine to cover the road ahead in healthcare delivery. There have been several efforts and initiatives to address specific healthcare use cases by applying ICT but, due to design methodologies centered on particular requirements, consequent solutions lack flexibility, scalability and reusability.

In this book chapter, we have faced the application of ICT within the healthcare environment from comprehensive vision of the whole complexity and potential solutions and technologies to apply by means of a methodological approach based on ODP-RM. We have addressed the requirements of interoperability, reusability and scalability by conforming international standards and establishing a technology neutral approach to cover all the current and future requirements of scenarios and domains within the healthcare environment. As proof-of-concept, we have applied this methodology to the development and deployment of a pervasive care platform.

There is a need of describing the whole health landscape in order to develop systems able to cooperate, exchange information, and serve as building blocks of more complex solutions. Only a proper methodology will be able to address the growing concerns, i.e. ageing of population, prevalence of the chronic condition in extensive population groups, the change of social models and structure of family, the impact of migration and population movements in the outbreak of infectious diseases, the corresponding growth in health expenditure, etc. Later, the next step should be the establishment of evaluation and certification mechanisms of healthcare services and systems, creating a framework which allows compare solutions and guarantee that deployed systems conform to common internationally accepted guidelines and methodologies.

Finally, we have to remark that although more efforts from the scientific and technical points of view, to establish guidelines and directives to reach a consensus about the proper application of ICT in a suitable future-proof way, only the policy makers, health stakeholders, and decision organisms will be able to make them effective.

Acknowledgments This work was supported in part by the Spanish National Board of Health Research (Instituto de Salud Carlos III), under grant PI082023, and by the General Directorate of Innovation, Science and Enterprising (Government of Andalucía), under grant P08-TIC-04069.

References

1. van Goor JN, Christensen JP, editors. Advances in medical informatics. Amsterdam: IOS Press; 1992.
2. Krol M. Telemedicine. IEEE Potentials 1997;16(4):29–31.
3. Laxminarayan SN. Editorial: information technology in biomedicine: maturational insights. IEEE Trans Inf Technol Biomed 2002;6(1):1–7.
4. Georgia Institute of Technology. Aware home [online]. 2000. http://awarehome.imtc.gatech.edu/. Accessed May 2010.
5. University of Illinois at Urbana-Champaign, Assisted Living Project. I-living [online]. http://lion.cs.uiuc.edu/assistedliving. Accessed May 2010.
6. Ambient Intelligence for the Networked Home Environment. Amigo [online]. 2008. http://www.hitech-projects.com/euprojects/amigo/. Accessed May 2010.
7. Saranummi N, Korhonen I, Kivisaari S, Ahjopalo H. A framework for developing distributed ICT applications for health distributed diagnosis and home healthcare. Proceedings of the 1st Transdisciplinary Conference on D2H2; Arlington, VA; 2006. p. 137–43.
8. OMG. Model-driven architecture [online]. 1997. http://www.omg.org/mda. Accessed May 2010
9. Ortwat C, Graefe A, Faulwasser T. Towards pervasive computing in health care—a literature review. BMC Med Inform Decis Mak 2008;8:26.
10. Martin H, Bernardos AM, Bergesio L, Tarrio P. Analysis of key aspects to manage wireless sensor networks in ambient assisted living environments. Proceedings of the 2nd International Symposium on Applied Sciences in Biomedical and Communication Technologies; 2009.
11. Sun H, Florio VD. Promises and challenges of ambient assisted living systems. Proceedings of the 6th International Conference on Information Technology: New Generations; 2009. p. 1201–7.
12. Estudillo-Valderrama MA, Roa-Romero LM, Reina-Tosina LJ, Naranjo-Hernández D. Distributed processing methodology for biomedical sensor networks: a novel approach. Proceedings of the 9th International Conference on Information Technology and Applications in Biomedicine; 2009. p. 1–4.
13. European Committee for Standardization CEN/TC 251. EN13940-1: health informatics—system of concepts to support continuity of care—part 1: basic concepts; 2006.
14. Spackman K, Campbell K, Côté R. SNOMED RT: a reference terminology for health care. Proceedings of the AMIA Annual Fall Symposium; 1997. p. 640–4.
15. Rector A, Nowlan W. The GALEN project. Comput Methods Programs Biomed. 1994; 45(1–2):75–8.
16. McDonald C, Huff S, Suico J, Hill G. LOINC, a universal standard for identifying laboratory observations: a 5 year update. Clin Chem. 2003;49:624–33.
17. Alonso G, Casati F, Kuno H, Machiraju V. Web services: concepts, architectures and applications. Heidelberg: Springer; 2004.
18. Foster I, Kesselman C. The grid 2: blueprint for a new computing infrastructure. San Francisco: Morgan Kaufmann Publishers, Inc.; 2003.
19. Buyya R, Yeo CS, Venugopal S, Broberg J, Brandic, I. Cloud computing and emerging IT platforms: vision, hype, and reality for delivering computing as the 5th utility. Future Generation Comput Syst 2009;25(6):599–616.

20. Roman I, Calvillo J, Rivas S, Roa L. Privilege management infrastructure for virtual organizations in healthcare grids. Proceedings of the 9th International Conference on Information Technology and Applications in Biomedicine; 2009. p. 1–4.
21. Augusto JC, Nakashima H, Aghajan H. Ambient intelligence and smart environments: a state of the art. Handbook of ambient intelligence and smart environments: part 1. New York: Springer; 2009. p. 3–31.
22. Papadopoulos A, Fotiadis DI, Lawo M. CHRONIOUS: a wearable system for the management of chronic disease patients. Proceedings of the 9th International Conference on Information Technology and Applications in Biomedicine; 2009. p. 1–4.
23. Nee O, Gorath T, Hülsmann N. SAPHIRE: intelligent healthcare monitoring based on semantic interoperability platform: pilot applications. Telemed E-Health Commun Syst 2008;2(2):192–201.
24. Estudillo-Valderrama MA, Roa-Romero LM, Reina-Tosina LJ, Naranjo-Hernández D. Design and implementation of a distributed fall detection system–personal server. IEEE Trans Inf Technol Biomed 2009;13(6):874–81.

Collective Health Intelligence: A Tool for Public Health

Andy Marsh, Denis Carroll and Richard Foggie

Abstract Web 3.0 is fast approaching. The European Union Future Internet Assembly, the roadmap for the Web heading towards semantic interoperability and building on the UK's adoption of the Internet and social media are accelerating this development. A number of health portals are opening, some with facilities for the capture of Patient Based Records. Collective Intelligence will be generated that, applied to health, has potential to support Public Health policy. By using the Internet, millions of people in the course of their daily activities contribute to uncertified data stores, some explicitly collaborating to create collective knowledge bases, some contributing implicitly through the patterns of their choices and actions. An application of soft computing, called Collective Health Intelligence, that reasons uncertified and certified data could enhance the social pool of existing health knowledge available to the public health agencies. Collective Health Intelligence could be used to complement national programmes by employing innovative sampling techniques, cost-effectively generating anonymous data trends that would quantify policy, indicate epidemiological effects and supply metrics to test policy efficacy.

Keywords Collective Intelligence · Soft Computing · Wisdom of the crowds · Web 2.0

A. Marsh (✉)
HoIP CIC, The Old Stable, Awbridge, Romsey, Hampshire SO51 0HN, UK
e-mail: am@hoip.eu

D. Carroll
University of Westminster, KTP Unit, 101 New Cavendish Street, London W1W 6XH, UK

R. Foggie
UK Department for Business, Innovation and Skills, 1 Victoria Street, London SW1H 0ET, UK

Commun Med Care Compunetics (2011) 1: 21–41
DOI: 10.1007/8754_2010_7

Published Online: 16 November 2010

1 Introduction

By using the Internet two new forms of information stores are being created in real time by thousands of people in the course of their daily activities, some explicitly collaborating to create collective knowledge bases like the Wikipedia [1] and Freebase [2], some contributing implicitly through the patterns of their choices and actions. Explicit knowledge bases, such as wiki's refine knowledge through the contributions of thousands of authors; implicit data stores allow the discovery of entirely new knowledge by capturing trillions of key clicks and decisions as people use the network in the course of their everyday lives.

The Internet plays a central role in providing health information. Increasingly, consumers are relying on the Internet for help with their health care decisions. According to Pew Internet [3], some 75–80% of Americans have used the Internet to find health information. As of January 2008, as reported by iCrossing [4], the Internet rivalled physicians as the leading source for health information. Using Web 2.0 tools, the Internet is becoming a platform for convening social networks and when patients managing the same chronic condition share observations with each other, their collective wisdom can yield clinical insights well beyond the understanding of any single patient or physician. The knowledge embedded and recorded within these societies or large groups of individuals is a form of Collective Intelligence [5].

Discovering and harnessing the tacit intelligence that results from the data generated by the activities of many people over time—revealed through analyses of patterns, correlations, and flows—is enabling ever more accurate predictions about people's preferences and behaviours, and helping researchers and everyday users understand and map relationships, and gauge the relative significance of ideas and events. Examples of uses for this type of intelligence already exist in industry. Google's PageRank system [6], which assigns value to a Web page based on the number of other pages that link to it, uses patterns discovered in hundreds of millions of links to determine which Web pages are most likely to be relevant in a list of search results. Amazon.com examines patterns in hundreds of buyer variables to recommend purchases that you might like based on your previous purchases, those of your friends, and other people who may have similar tastes or preferences.

Scientists at Cornell [7] have created a machine that can work out the laws of nature by observing the world around it. The machine uses a computer program that can search through huge amounts of data and look for underlying patterns. The machine took only hours to come up with the basic laws of motion. Moreover, to prepare scientists for an era when machines become cleverer than people, Google and NASA are throwing their weight behind a new school for futurists in Silicon Valley, called the Singularity University [8], that will offer courses including Artificial Intelligence. Their backing demonstrates the growing mainstream acceptance that before the middle of this century artificial intelligence will outstrip human beings, ushering in a new era of civilisation.

After setting the policy context in the next section, the third and fourth section address from where data originates both in the context of certified data feeds from evidence-based resources and uncertified data feeds from resources such social media which encapsulate the wisdom of the crowds. The use of soft computing to support Collective Health Intelligence is presented in the fifth section. Its application to support public health policy is presented in the sixth section. Conclusions and recommendations and presented in the seventh section.

2 Policy Context

The first web was predicated on governanced, certified information repositories being accessed by individual users. Web 2.0 has come into being through harnessing user generating content. This can be made compelling by the provision of Application Programming Interfaces (APIs) that make it easy for users to contribute content into communities. Significant, successful commercial and social sites now exist (e.g. Facebook, e-Bay) and a similar evolution is occurring among smart phone users.

By 2012, everyone in the UK should have access to a 2 Mbps Internet connection as outlined in the Digital Britain interim report [9], published June 2009. The report presents the key objectives for shifting Britain to the forefront of the digital economy with a 22-point action plan that will help achieve five priorities which government has identified, namely (1) creating a dynamic investment climate, (2) preserving UK content, (3) improving digital literacy, (4) upgrading and modernising digital infrastructure and (5) developing the infrastructure, skills and take-up. Digital Britain outputs will underpin the growth of those opportunities by increasing the depth and range of connectivity across the UK. Development of seamless connectivity across popular platforms (mobile phones and interactive TV) will provide an infrastructure that assists the 'Putting People First' [10] concordat, allowing delivery of personalised, transformed social care.

To protect the public from threats to their health from infectious diseases and environmental hazards, the Health Protection Agency (HPA) [11], an independent UK organization, was set up by government in 2003. It does this by providing advice and information to the general public, to health professionals such as doctors and nurses, and to national and local government. As defined in its research and development strategy 2005–2010 [12], one of the main research themes (Theme 14) is *surveillance development and data analysis* which states that rapid advances are taking place in computations sciences, data storage and retrieval, and data analysis. As reported by IBM [13], it is essential that all public health data collection and the result of their analysis are fit for purpose, available rapidly, and available to others for further analysis in an interactive way. The HPA will ensure that it is informed of all the new development in information and data sciences to allow their rapid assessment for use for health protection purposes.

The UK Government White Paper 'Choosing health' [14] recognised the need for good information and knowledge management. The Department of Health (DH) [15] subsequently published a national information and intelligence strategy called 'Informing healthier choices' [16] that addresses availability, timeliness and quality of health information and intelligence across England as well as knowledge management and support for those working to improve and protect public health. The strategy aims to improve the amount and quality of data on contemporary health challenges being collected and made available. Prevalence models will be generated for common health problems that commissioners will need to address. This will allow them to estimate current and future disease burdens as well as assessing the impact of their interventions. The strategy will also support professionals working in public health by designing health information and intelligence systems at all levels.

The next two sections address from where information/data originates both in the context of certified data feeds from evidence-based resources and uncertified data feeds from resources such as social media which encapsulate the wisdom of the crowds.

3 Certified Data Feeds from Evidence-Based Resources

The systematic review of published research studies is a major method used for evaluating and assessing the quality of evidence relevant to the risks and benefits of treatments (including lack of treatment). The Cochrane Collaboration [17] is one of the most well known and well respected examples of systematic reviews. Others include Bandolier [18], Cochrane Library Database of Systematic Reviews [19], The Database of Abstracts of Reviews of Effects (DARE) [20], Health Technology Assessment Database (HTA) [21], NHS Economic Evaluation Database (NHS EED) [22] and the UK Database of Uncertainties about the Effects of Treatments [23].

The Specialist Libraries of the National Library for Health [24] filter the huge quantity of published research into a trusted, relevant, comprehensive library of the latest guidelines, systematic reviews and specialized areas of interest. The majority of NLH Specialist Libraries focus on particular health problems, such as cancer [25] and cardiovascular [26]. A few NLH Specialist Libraries focus on specific patient groups, such as child health [27] and the focus of other NLH Specialist Libraries is on aspects of health services, such as commissioning [28].

The National Library of Guidelines [29](formerly known as the Guidelines Finder), which also includes guidelines from the National Institute for Health and Clinical Excellence (NICE) [30], indexes over 2000 selected care guidelines available online in full text format. Coverage is predominantly UK with some additional international guidelines from the US National Guidelines Clearing House [31], New Zealand Guidelines Group—Guidelines Library [32] and Australian National Health and Medical Research Council: Clinical Practice

Guidelines [33]. Additionally, Clinical Knowledge Summaries [34] provide an up-to-date source of clinical knowledge, incorporating PRODIGY resources, that can help healthcare professional and patients in managing the common conditions generally seen in primary and first-contact care.

The policy that requires researchers funded by the United States National Institutes of Health (NIH) [35] to deposit their peer-reviewed research articles into PubMed Central (PMC) [36], the free online digital archive of full-text, peer reviewed research publications, was made permanent, on 12th March 2009, by President Obama (through a provision written into the 2009 Consolidated Appropriations Act) [37]. The NIH Public Access Policy [38] ensures that the public has access to the published results of NIH funded research. It requires scientists to submit final peer-reviewed journal manuscripts that arise from NIH funds to the digital archive PubMed Central *upon acceptance for publication*. The Policy requires that these papers are accessible to the public on PubMed Central no later than 12 months after publication. In the UK, the organisations that fund UK PubMed Central [39], such as the Medical Research Council [40] stipulate that published research, arising from the research grants they award, must be made available through the UK PubMed Central repository, typically within 6 months of being published. In addition to PubMed the National Library for Health provides access to a number of Healthcare Databases [41] including AMED, British Nursing Index, CINAHL, E-books, EMBASE, HMIC, MEDLINE, My Journals and PsycINFO.

In the UK, the Association of Public Health Observatories (APHO) [42] represents and co-ordinates a network of 12 public health observatories (PHOs) working across the five nations of England, Scotland, Wales, Northern Ireland and the Republic of Ireland. The network of PHOs produce information, data and intelligence on people's health and health care for practitioners, policy makers and the wider community. The key role of APHO is to develop public health expertise and in-depth knowledge of health and health care at a regional level. APHO monitors and forecasts trends in health status and disease, and shows how health inequalities are being tackled locally and regionally. Furthermore APHO monitors the effects of health and health care interventions to help give commissioners and providers of health and related services scientific evidence and data to reduce inequalities in access and outcomes. The APHO have been commissioned by the Department of Health as a deliverable of Informing Healthier Choices: Information and Intelligence for Healthier Populations strategy to develop a new National Library for Public Health. The project is being delivered by the North East Public Health Observatory (NEPHO) [43] on behalf of APHO.

The National Library for Public Health (NLPH) [44] is a specialist library of the National Library for Health (NLH) [45] and contains resources in line with the objectives of the white papers *Choosing Health: making healthy choices easier* and *Our health, our care, our say*. The purpose of the NLPH is to promote and enable the use of evidence based practice in all aspects of population health. The library is both a collection of knowledge resources, and a community of practice serving the overlapping domains of public health and population-based commissioning. The library includes population health related resources contained within the NLH "core content" produced, commissioned or licensed by the NHS in the United Kingdom. These include NICE guidelines [46], Cochrane systematic reviews [19], and Clinical knowledge summaries (CKS) chapters [47]. The library also indexes patient information, critically appraised resources within health and social sciences and relevant policy documents. Comprehensive searches for evidence based guideline relating to population health and systematic literature searches for key topic areas for population health are regularly updated. Resources that have been retrieved from the NLH core content and other Specialist Libraries which satisfy relevant NLPH quality control criteria are indexed where appropriate using the high level taxonomy taken from the Public Health Language project [48] (sub-set of SNOMED-CT) and entered.

With the introduction of Web 2.0 technologies, social networking environments have become a repository of public opinion. These uncertified data feeds are presented in the next section.

4 Uncertified Data Feeds—Wisdom of the Crowds

Web crawling or *spidering* uses a computer program called a Web crawler that browses the World Wide Web in a methodical, automated manner. Other terms for Web crawlers are *ants*, *automatic indexers*, *bots*, and *worms* or *Web spider*, *Web robot*, or *Web scutter*. Many sites, in particular search engines, use spidering as a means of providing up-to-date data. Web crawlers are mainly used to create a copy of all the visited pages for later processing by a search engine that will index the downloaded pages to provide fast searches. Crawlers can also be used for automating maintenance tasks on a Web site, such as checking links or validating HTML code. Also, crawlers can be used to gather specific types of information from Web pages, such as harvesting e-mail addresses. In general, a web crawler starts with a list of URLs to visit, called the *seeds*. As the crawler visits these URLs, it identifies all the hyperlinks in the page and adds them to the list of URLs to visit, called the *crawl frontier*. URLs from the frontier are recursively visited according to a set of policies.

Web analytics is the measurement, collection, analysis and reporting of internet data for purposes of understanding and optimizing web usage. There are two categories of web analytics; *off-site* and *on-site* web analytics. Off-site web analytics tools measure the potential of a websites audience. There are two types of techniques that achieve this—using panel data or Internet Service Provider (ISP) data. Companies such as comScore [49] and Nielsen Netratings [50] use the panel method by recruiting participants to have monitoring software installed on their computers to measure their web activity. Alternatively, companies such as Hitwise [51](now part of Experian), collect off-site visitor information by aggregating anonymous data provided by ISPs. On-site web analytics measure a visitor's journey once on a website. There are two main technological approaches to collecting the data. The first method, *log file analysis*, reads the log files in which the web server records all its transactions. The second method, *page tagging*, uses JavaScript on each page to notify a third-party server when a page is rendered by a web browser. Both collect data that can be processed to produce web traffic reports. Many different vendors provide on-site web analytics software and services, such as Google Analytics. A key difference between on-site and off-site web analytics tools is that on-site visitor data is only available to the website owner and the people he/she grants access to, such as a third-party marketing agency. Conversely off-site web analytics data can be obtained for any website—including your competitors and partners, provided there is sufficient visit data.

Another approach to collecting data is the somewhat controversial approach of Deep packet inspection (DPI) which involves examining both the data and the header of an information packet as it passes a 'black box' on a network, in order to reveal the content of the communication. Targeted advertising services, such as Phorm [52] in the UK, use DPI to monitor anonimized user behavior and to target adverts at those users. In addition, UK government initiatives such as the Intercept Modernisation Programme have proposed using DPI to perform mass surveillance of the web communications of the entire UK population. Webwise is name of system used by Phorm.

In recent years, the growth of Social Networking sites, such as Facebook, Twitter, MySpace, etc. has been phenomenal and it is still continuing to grow. Across the globe, the audience for social media services has risen 25% between August 2007 and August 2008, according to comScore. Facebook, is reportedly on track to having half of the UK population registered as members in the near future and the average user is said to use the site for a quite astounding 6 h a month. And this only looks set to increase, with the recent figures showing that, in the past year, the number of unique visits has risen by 107% from January 2008 to January 2009. The other Social Network contenders, MySpace and Bebo, are also continuing to expand, but at a fraction of the rate of the leader, Facebook.

Using social media prior to the G20 meeting in London [53], the Metropolitan Police analysed activity on websites including Facebook, MySpace and many others to see how many were likely to join the call for protests at the G20. Plans for how many policemen and women should be put on the streets during the G20 were drawn up with regard to this analysis. But it was during the 2-day event that social media, in particular Twitter, came into their own. Thousands of messages marked with a #G20 tag passed through Twitter and organisations such as IndyMedia [54] had their own tags that helped to collate information about what happened during the protests.

For Anjali Kwatra, a spokesperson for Action Aid, Twitter was essential in giving supporters outside a glimpse of what was happening inside the Excel centre. The immediacy of the medium also appealed, said Ms Kwatra, adding that the 140 character limit gave messages an informality over other ways of communicating. "We are also blogging, but that takes time and goes on to the Action Aid website, where people have to come in and look at it," said Ms Kwatra. "It's a mix of both the observational and informative," she said.

Dominic Casciani, from BBC News, who has been mixing with protesters across London, said many of the groups used Twitter as a way to reach out to supporters more quickly than they could with e-mail or text messages. "It's not the defining tool because it's never going to replace the mobile phone or text message to close friends," he said, "but there's clearly something in broadcasting to subsets of people that are looking for specific information."

In the space of just 1 year the Twitter traffic has increased by a whopping 974% in the UK alone, which is estimated as four to 6 million people now Twittering [55]. A comprehensive list of doctors, medical students and medicine related tweets and blogs/websites can be found at Medical Student Blog [56].

There are a range of social media sites established by individuals and organizations which provide spaces for groups to interact within the health context. Some of these are for professionals only, some are user-driven around specific conditions with clinicians as participants and some bring together users and clinicians for research and service provision. Blogs, chat rooms, discussion groups all provide space for individuals to interact and discuss. In doing so they may be choosing to share health information about themselves or about others. For instance clinicians may discuss the best way to treat swine-flu, others may ask for help on diagnosis; individuals may share that they are 'off-work' today because they have flu or a stomach-bug or that they are looking after a relative who is ill.

The web has provided access to data, and billions of statistics on every matter, that would never have found before, from specialist datasets to macroeconomic minutiae. It has also provided tools to visualize that information and mashing it up with different datasets. However, looking for the simplest fact or statistic and the search engines, such as Google, will present a million contradictory ones. Furthermore, official documents are often published as uneditable pdf files—useless for analysis except in ways already done by the publishing organization themselves. Alternatively, sometimes an avalanche of facts is unleashed in order to bury the truth. Journalists have to walk this tightrope every day, ensuring that the numbers that are published are correct. Increasingly reporters around the world are making it their mission to make data truly free; to publish everything. A leading example is the newly released "Data-store" from the UK's Guardian newspaper [57, 58] will publishes the raw statistics behind the news as a collection of data sets that are made open to the public. Another example is the UK NHS choices *Behind the headlines* [59] which provides an unbiased and evidence-based daily analysis of the science behind health stories that make the news.

The use of sophisticated internet search technologies (currently text based search but shortly to be supplemented by the ability to search audio and video feeds) provides the ability to identify and track emergent discussions and commentary around specific diseases and outbreaks. The most advanced example of this is HealthMap Global Disease Alert Map [60]. In its own words

> HealthMap brings together disparate data sources to achieve a unified and comprehensive view of the current global state of infectious diseases and their effect on human and animal health. This freely available Web site integrates outbreak data of varying reliability, ranging from news sources (such as Google News) to curated personal accounts

(such as ProMED) to validated official alerts (such as World Health Organization). Through an automated text processing system, the data is aggregated by disease and displayed by location for user-friendly access to the original alert. HealthMap provides a jumping-off point for real-time information on emerging infectious diseases and has particular interest for public health officials and international travelers.

A recently (21 May 2009) published paper in NEJM entitled 'Digital Disease Detection—Harnessing the Web for Public Health Surveillance' [61] provides an excellent overview of this area and a primary reference for activity in this field.

In the UK, the Office for National Statistics (ONS) [62] is the executive office of the UK Statistics Authority [63], a non-ministerial department which reports directly to Parliament. ONS is the UK Government's single largest statistical producer. It functions as: the office of the National Statistician, who is also the UK Statistics Authority's Chief Executive and principal statistical adviser; the UK's National Statistics Institute (or NSI—to use European terminology), and the 'Head Office' of the Government Statistical Service (GSS) [64]. The Health and Social Care theme [65] covers statistics relating to information gathered on public health, health services provided by the National Health Service (NHS), social care, and health and safety at work.

Currently communicable disease surveillance within England uses a network of 'spotter' GP practices submitting information as one channel for surveillance (Real-time Syndromic Surveillance) [66]. The outbreak of Swine Flu in Mexico City, in April 2009, highlights the issue that the automatic identification of notifiable diseases from electronic medical records, could potentially improve the timeliness and completeness of public health surveillance. A study published in 2008, Automated Identification of Acute Hepatitis B Using Electronic Medical Record Data to Facilitate Public Health Surveillance [67], conducted in part by the Massachusetts Department of Public Health, and the Harvard Medical School, concluded, "electronic medical record data can reliably detect acute hepatitis B. The completeness of public health surveillance may be improved by automatically identifying notifiable diseases from electronic medical record data."

Furthermore, the linkage to other information such as prescribing data and past medical conditions would allow the development of risk-stratification of individuals and the subsequent provision of individual tailored alerts through for example HealthSpace. An example would be reminders for flu-vaccinations. The US Centers for Disease Control (CDC) National Centre for Public Health Informatics is currently working with GE Healthcare to integrate public health alerts into the locally available records systems as well as the ability to highlight patients most at risk or whose clinical profile seems to fit the relevant alert condition [68].

5 Collective Health Intelligence

As the Internet has developed, so has the concept of Collective intelligence [69] as a shared public forum. Web 2.0 has enabled interactivity and thus, users are able to generate their own content. Collective intelligence is a shared or group intelligence that emerges from the collaboration and competition of many individuals. The global accessibility and availability of the Internet has allowed more people than ever to contribute their ideas and to access these collaborative intelligence spaces [70].

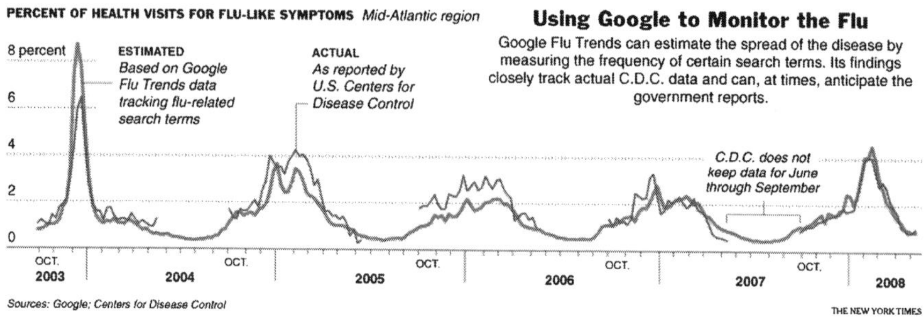

In moving further into the age of machine intelligence and automated reasoning the Web and especially search engines, such as Google, have reached a point where systems have a high machine IQ (MIQ) [71] or in the context of the Web, Web IQ (WIQ) [72]. The data collected by search engines is particularly powerful, because the keywords and phrases that people type into them represent their most immediate intentions. Analysis of the sheer volumes of information available and the ability to differentiate signal trends from the 'background noise' of the internet has significant potential. Tests of the new Web tool from Google.org suggest that it may be able to detect regional outbreaks of the flu, for the US, a week to 10 days before they are reported by the Centers for Disease Control (CDC) [73]. It turns out that a lot of ailing Americans enter phrases like "flu symptoms" into Google and other search engines before they call their doctors. To develop the service, Google's engineers devised a basket of keywords and phrases related to the flu, including thermometer, flu symptoms, muscle aches and chest congestion. Google then dug into its database, extracted 5 years of data on those queries and mapped it onto the CDC's reports of influenza like illness. Google found a strong correlation between its data and the reports. Google Flu Trends avoids privacy pitfalls by relying only on aggregated data that cannot be traced to individual searchers. Its Google Flu Trends service [74] analyzes those searches as they come in, creating graphs and maps of the country that, ideally, show where the flu is spreading. The CDC reports are slower because they rely on data collected and compiled from thousands of health care providers, labs and other sources. The premise behind Google Flu Trends—a fruitful marriage of mob behavior and medicine, a form of "collective intelligence"—has been validated by an unrelated study indicating that the data collected by Yahoo, can also help with early detection of the flu.

Existing search engines have many remarkable capabilities. However, what is not among them is a deduction capability—the capability to answer a query by a synthesis of information which resides in various parts of the knowledge base.

A question-answering system is by definition a system which has this capability. One of the principal goals of Web intelligence is that of upgrading search engines to question-answering systems [75]. Many modern machine learning, or automated reasoning, methods such as the Semantic Web, Cyc, OWL [76–78] and other ontology-centered systems, such as Open Clinical [79], are based on objectivist Bayesian principles—bivalent logic and bivalent-logic based probability theory [80].

One of the crucial features of the Bayesian view is that a probability can be assigned to a hypothesis where a hypothesis can only be rejected or not rejected. One leading example of the application of Bayesian inference is by Autonomy Corporation PLC [81], an enterprise software company with joint head quarters in Cambridge, UK, and San Francisco, USA. It has developed a variety of enterprise search and knowledge management applications using adaptive pattern recognition techniques centered on Bayesian inference [82] (statistical inference in which evidence or observations are used to update or to newly infer the probability that a hypothesis may be true) in conjunction with traditional methods, such as information theory [83].

Bivalent-logic-based methods of knowledge representation and deduction are of limited effectiveness in dealing with information which is imprecise or partially true [72]. Much of the information which resides in the Web—and especially in the domain of world knowledge—is imprecise, uncertain and partially true [84, 85]. This information can be subdivided into *measurement-based* which can be represented numerically (e.g. It is 35°C, Eva is 28, Tandy is 3 years older than Dana) or *perception-based* which can be represented linguistically (e.g. It is very warm, Eva is young, Tandy is a few years older than Dana, It is cloudy, Traffic is heavy). More specifically, perceptions are f-granular in the sense that (a) the boundaries of perceived classes are unsharp; and (b) the values of perceived attributes are granular, with a granule being a clump of values drawn together by indistinguishability, similarity, proximity or functionality [86].

For example, designing an expert system to mimic the diagnostic powers of a physician, one of the major tasks to codify is the physician's decision-making process. The physician's view of the world, despite their dependence upon precise, scientific tests and measurements, incorporates evaluations of symptoms, and relationships between them, in a "fuzzy," intuitive manner: deciding how much of a particular medication to administer will have as much to do with the physician's sense of the relative "strength" of the patient's symptoms as it will their height/weight ratio. While some of the decisions and calculations could be done using Bayesian inference an alternative is fuzzy logic. Fuzzy logic is an extension of boolean logic dealing with the concept of *partial truth*. It is related to fuzzy sets and possibility theory and was introduced in 1965 by Dr. Lotfi Zadeh of Berkeley [87]. Whereas classical logic holds that everything can be expressed in binary terms (0 or 1, black or white, yes or no), fuzzy logic replaces boolean truth values with degrees of truth. Bart Kosko [88] has shown that probability is a subtheory of fuzzy logic, as probability only handles one kind of uncertainty.

Fuzzy logic is best applied to perception-based information, whereas Bayesian inference, or probabilistic reasoning, is best applied to measurement-based information. In many cases, especially in health, the available information is a mixture of both and can be interpreted most effectively by using a combination of fuzzy logic and probabilistic reasoning subsuming belief networks, evolutionary computing including DNA computing, chaos theory and parts of learning theory. Referred to as *approximate reasoning*, this hybrid system can also be supported by *search & optimization* techniques, including neural networks, machine learning and evolutionary algorithms. Collectively, these represent the main components of Soft Computing [89] and when applied to health, termed *Collective Health Intelligence,* can be used to enhance the social pool of existing health knowledge available to public health agencies.

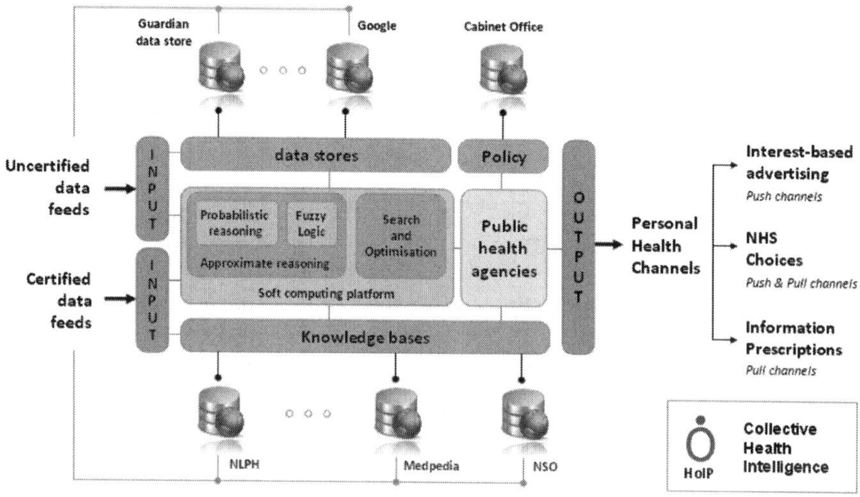

Within a soft computing platform approximate reasoning can be combined with search and optimisation techniques to analyse uncertified and certified data feeds in conjunction with their respective stores. Uncertified data originates from the thousands of people using the Internet in the course of their daily activities, some explicitly collaborating to create collective knowledge bases, some contributing implicitly through the patterns of their choices and actions. Certified data feeds represent *measurement-based* information and include Evidence-based reviews, Specialist Libraries, Guidelines, Research publications, information from the APHO and knowledge in the NLPH. Resulting collective intelligence can be used by public health agencies to support public health policy and disseminated over personal health channels.

Collective Health Intelligence can be used to extract knowledge and fine-tune policy but there still needs to be a mechanism to disseminate this information. As the next section presents this can be achieved using a number of approaches either pushing or pulling information or combinations of both.

5.1 Personalised Health Channels

Google launched, on Mar 11th 2009, a form of behavioral targeting advertising named Interest-Based Advertising [90] that will allow users to decide which ads they want to see as they surf the web. As a form of push health channels users can create a profile of their interests or wait for Google to customize them automatically based on the users browsing habits. Interest based categories are based on the type of web site a browser visits. For example, if a user visits ESPN often, Google will know that the user is interested in sports. Currently there are some 30 top line categories and about 600 detailed categories. In addition, users have control over these categories and can add or remove categories in the user ad preferences section. Previous interaction is the second area of Interest Based advertising where Google, leveraging their DoubleClick technology [91], is able to show ads to users based on their browsers previous interaction with that advertiser. For example, if a user had a product in their shopping cart and did not check out, the advertiser can display ads on other sites, within the Google network, that promote that product or that product line. With a larger launch planned for later on in the year, initially those advertisers included in this beta will have a special portal to manage their ads. But ultimately, it is Google's goal to build the solution directly into the AdWords console.

As an example of an alternative pull health channel, the NHS information prescription project [91], launched in September 2008, offers everyone who has a long-term condition or social care need an 'information prescription', in consultation with a health or social care professional. Information prescriptions let people know where to get advice, where to get support and where to network with others with a similar condition. They will include addresses, telephone numbers and website addresses that people may find helpful, and show where they can go to find out more. They will help people to access information when they need it and in the ways that they prefer. An Information Prescription has five main components: *Information content*—the identification of reliable and relevant sources of information. *Directories*—repositories of information that link to individual Information Prescriptions. *Personalised process*—information is provided that is specific to the condition, place and point on the care pathway. *Issuing or prescribing*—creating and offering an Information Prescription to a user or carer. *Access*—Information Prescriptions are made available to users through a range of accessible channels, such as face-to-face engagement, the Internet, email, telephone and outreach.

By combining push and pull health channels, the National Health Service (NHS) website, called NHS Choices, now includes a number of Web 2.0 tools including wikis, blogs, RSS feeds and social media tools. The NHS Choices blogs [92] lets users read, and comment on, the latest views, news and tips from more than 100 patients, carers and medical professionals writing across currently nine topics. Currently the NHS choices blogs address arthritis, asthma, COPD, depression, diabetes, heart conditions, kidney disease, prostate cancer and pregnancy. This service which will soon include a range of discussion forums and an

"Ask the doctor" feature where GPs will select the most interesting questions and publish the answers. In addition, NHS Choices also offers a number of RSS feeds and every page from *Live well*, *Health A-Z* and *behind the headlines* can be included as links in a range of social bookmarking websites including Delicious, Digg, Facebook, Reddit, StumbleUpon. NHS Choices also supports a mobile text service with DirectGov [93] that enables a user to find local NHS services, such as hospitals or dentists. It also offers several interactive tools, for example, you can check your alcohol intake, the costs of smoking or see if you are healthy weight. Furthermore, NHS Choices main, NHS Choices blogs and NHS Direct are amongst the NHS organisations on Twitter. Furthermore, the entire NHS library is to be part of the recently launched (Feb 2009) wiki site called Medpedia [94]. As Medpedia grows over the next few years, it will become a repository of up-to-date unbiased medical information, contributed and maintained by health experts around the world, and freely available to everyone.

As an example of personal health portal, the NHS HealthSpace [95] offers registered users an on-line facility to record health and lifestyle data. This should be a rich source of 'semi-certified' data provided uptake is encouraged. The portal is ideally suited to deliver targeted rich content should the use of Collective Health Intelligence indicate a need to advise or alert. Other Patient Based Records systems are available, notably Microsoft HealthVault™ [96] and Google Health (currently a beta release) [97]. The development of Patient Based Records and Personal Health Portals will transform the relationship and modes of interaction with health professionals, potentially to the advantage of all. A note of caution; in April 2009 the Boston Globe reported that a US man who imported his medical records into Google Health was surprised to learn of spurious major health concerns [98]. It appears that this transpired because of poor data coding on insurance records.

As presented in the next section, Collective Health Intelligence can be used to fine-tune public health policy.

6 Collective Health Intelligence Support for Public Health Policy

Reducing obesity is also one of the six overarching priorities in the public health white paper Choosing Health [99]. Obesity develops from an accumulation of excess body fat, which occurs when energy intake from food and drink consumption is greater than energy expenditure through the body's metabolism and physical activity. However the causes of obesity are more complex than this, and relate to a wide variety of societal and behavioral factors.

The average British Body Mass Index (BMI) has increased for men from 26 to 27.3 whilst for women from 25.8 to 26.9. Obesity is also associated with health problems which include type 2 diabetes, cardiovascular disease and cancer. The resulting NHS costs attributable to overweight and obesity are projected to reach

£9.7 billion by 2050, with wider costs to society estimated to reach £49.9 billion per year (Foresight 2007) [100]. These factors combine to make the prevention of obesity a major public health challenge.

For the UK 30,000 deaths can be attributed to obesity and some 18 million sick days, this condition is estimated to cost the NHS about £500,000 a year in treatment costs. One in six PCTs have increased spending to be able to treat obesity as well as purchase of dedicated equipment for obese patients. This is the motivation for GlaxoSmithKline to release its anti-obesity drug Alli.

The new National Obesity Observatory [101] for England was established to provide a single point of contact for wide-ranging authoritative information on data, evidence and practice related to obesity, overweight, underweight and their determinants. This specialist observatory is a member of the Association of Public Health Observatories [42] and is sited alongside the South East Public Health Observatory [102]. The National Obesity Observatory will work closely with a wide range of organizations and will provide support to policy makers and practitioners involved in obesity and related issues.

As reported on the National Obesity Observatory website there has been a rapid increase in the prevalence of overweight and obesity in recent years, with the proportion of adults in England with a healthy BMI (18.5–24.9) decreasing between 1993 and 2007 from 41 to 34% among men and 50 to 42% among women. Currently 24% of men and women (aged 16 years and over) are obese (HSE 2007). In adults, obesity is commonly defined by a Body Mass Index (BMI) of 30 or more. In addition 10.4% of boys and 8.8% of girls (average 9.6%) in Reception year (aged 4–5 years) and 20% of boys and 16.6% of girls (average 18.3%) in Year 6 (aged 10–11 years) are also classified as obese according to the British 1990 population monitoring definition of obesity (≥95th centile) [103]. By 2050 the prevalence of obesity is predicted to affect 60% of adult men, 50% of adult women and 25% of children (Foresight 2007).

Six new reviews examined the effect of physical activity in children. The most recent Cochrane review [104] examining school based physical activity programs for promoting physical activity and fitness in young people aged 6–18, overall found no significant impact on BMI, despite demonstrable positive effect on other lifestyle behaviors (such as reduced TV viewing, an increase in duration of activity, and improved VO2 max and blood cholesterol). These findings were supported by Benson et al. [105], who was unable to identify a significant impact of BMI after reviewing the effect of resistance training; and Kelley et al. [106] who found no significant impact on BMI, after examining studies evaluating the impact of aerobic exercise in children, despite observing significant improvements in percentage body fat. Three further reviews addressed active travel to school, providing evidence to suggest that: (1) parental and child attitudes to active travel is affected by fears over safety, perception of risk, dislike of the local environment and cultural preference for car usage, despite more young people showing a desire to walk and cycle more (Lorenc et al. [107]); (2) active travel is associated with increased activity levels in young people, although the relationship with weight status is far less clear [108, 109].

Four new reviews examined the impact of specific dietary components. Three of these reviews examined the impact of sugar sweetened soft drinks. Although Wolff et al. [110] identified observational studies that support the hypothesis linking sugar sweetened soft drinks to weight gain, all three reviews (Gibson [111] and Forshee et al. [112] both received funding from the soft drinks industry) reported inconsistencies between included studies, and were therefore unable to identify any significant relationship. A review by Rosenheck [113] identified a relationship between fast food consumption and weight gain, however much of this evidence derived from adult populations, with more research required for children.

The food environment and policy was reviewed in three papers. Holsten [114] examined the current evidence base surrounding the relationship between obesity and the community food environment. However this research area remains at an early stage, with many of the included studies subject to methodological limitations, and very few taking place within child populations. Jaime and Lock [115] examined the impact of school based food and nutrition policies and obesity. Their findings demonstrated that few large scale or national policies have been evaluated. For those school food policies that have been evaluated, some were able to show efficacy in improving the food environment and dietary intake within school, although evidence on their impact on BMI was very limited. A further review of school fruit and vegetable programmes was undertaken by de Sa and Lock [116]. Whilst the authors found evidence to suggest that school fruit and vegetables schemes can be effective at increasing fruit and vegetable intake and can contribute to the reduction in diet inequalities, there was very little evidence examining the impact on child weight status.

A meta-analysis of longitudinal studies of depression and weight control was carried out by Blaine [117]. The findings demonstrate a significant correlation between depression and later obesity, particularly for adolescent females, thus highlighting the importance of depression screening and treatment in obesity prevention programmes.

In conclusion, the current update demonstrates of the growth of evidence within this field, with reviews now covering broader issues around the environment, policy and practice. Although the systematic reviews vary in their quality assurance and inclusion criteria, collectively the available evidence base remains weak, due to the inconsistencies and heterogeneity of available studies. To a certain extent this is indicative of the complexity of the obesity however, the evidence from these systematic reviews will continue to build a foundation for future research and help to inform practice and policy.

Independent charity organizations such as the Child Growth Foundation (CGF) [118] and National Obesity Forum (NOF) [119] recommend the enacting of provisions made in the Health & Social Care Act 2008 [120] which foresees annual checks throughout the UK from the age of 2. NOF states, in its Recommendations for primary prevention [121] that regular and accurate measurement of children should be the first step in the prevention of obesity and they would like to see further body mass index (BMI) assessment in schools. The Reception Year and Yr

6 measures are good public health *snapshots* of the nation's childhood epidemic but fail completely to identify any escalation of unhealthy weight in school.

The expansion of existing surveillance schemes was also identified by the Foresight project "Tackling obesities : Future Choices" [100] as one of a number of recurring themes to facilitate obesity research which also included (a) Exploitation of existing datasets in public and private domain; (b) Improvements in the methodologies to measure human behavior, especially diet and physical activity; (c) Development of more detailed models to examine the future impact of obesity and its comorbidities including the use of microsimulation techniques.

Collective Health Intelligence has the potential to address some of the surveillance challenges such as ability to collect objective data in non-intrusive ways at scale and low cost, particularly data which would be unobtainable through traditional channels (e.g. the surveillance of BMI beyond the NCMP). For example, Internet connected weighing scales automatically uploading their data to online servers for user browsing is now becoming readily obtainable in the consumer market. Collectively, if made available, this uncertified data provides an additional means of whole population measurement/screening that when combined with certified data could be used to support large-scale evaluation of public health policy related to obesity linking trials and best practices.

7 Conclusions

Data is the most basic building block of any modern health information infrastructure. Addressing the *surveillance development and data analysis* theme of the HPA research and development strategy [12] and in accordance with government policy [122, 123] one key challenge is to establish a task force to assess the accessibility and interoperability of inter-agency data, information and knowledge that when used by Collective Health Intelligence can directly or indirectly influence public health policy.

By using the Internet two new forms of information stores are being created in real time by thousands of people in the course of their daily activities, some explicitly collaborating to create collective knowledge bases, some contributing implicitly through the patterns of their choices and actions. A key challenge is to develop an open platform in accordance with government policy: Open Source, Open Standards and Re-Use: Government Action Plan [124], for Collective Health Intelligence, with effective surveillance and reasoning capabilities, to conduct complex analyses of these voluminous information stores and assess their impact on public health policy.

One aim of the Department of Health (DH) national information and intelligence strategy called 'Informing healthier choices' [16] is to improve the availability, timeliness and quality of health information across England. A key challenge for Collective Health Intelligence, equal to the NHS information prescription project [91], is to utilise unified communications and social media technologies to promote personalised health channels that will deliver intelligence at the right time ensuring that information is tailored around the specific needs of users when they come into contact with service providers, in the most appropriate way ensuring that information is provided through a range of channels and is accessible and convenient, in the most appropriate format working to develop a range of formats to maximise the inclusiveness of the Information Prescription process, helping to make sure that information is accessible and useable for all and be available at the right place ensuring information is made available from a range of access points, situated at locations convenient for the service user.

The success of Collective Health Intelligence depends on the use of technologies that provide timely and accurate measurement-based and perception-based information. There must be an infrastructure for information collection, exchange and analysis with interoperability that facilitates information sharing between organizations, regions, and agencies. A key challenge for Collective Health Intelligence would be to demonstrate its effectiveness and potential impact on public health policy by addressing one of the six overarching priorities in the public health white paper Choosing Health [125] such as reducing obesity and more specifically childhood obesity.

Acknowledgments The authors would like to acknowledge the Public Health, Research and Development Department of Health and the i4i Programme of the National Institutes for Health Research in the UK for financially supporting the research scoping study and background material for this chapter.

References

1. http://en.wikipedia.org.
2. http://www.freebase.com.
3. http://www.pewinternet.org/Reports/2008/The-Engaged-Epatient-Population.aspx.
4. http://www.icrossing.com/research/how-america-searches-health-and-wellness.php.
5. http://en.wikipedia.org/wiki/Collective_Intelligence.
6. http://www.google.com/corporate/tech.html.
7. http://www.sciencemag.org/cgi/content/abstract/sci;324/5923/81.
8. http://singularityu.org/.
9. http://www.culture.gov.uk/images/publications/digital_britain_interimreportjan09.pdf.
10. http://www.cpa.org.uk/cpa/putting_people_first.pdf.
11. http://www.hpa.org.uk/.
12. http://www.hpa.org.uk/web/HPAweb&HPAwebStandard/HPAweb_C/1203064747700.
13. http://whitepapers.silicon.com/0,39024759,60243863p,00.htm?s_cid=38.
14. http://www.dh.gov.uk/en/Consultations/Closedconsultations/DH_4084418.
15. http://www.dh.gov.uk/.
16. http://www.dh.gov.uk/en/Consultations/Closedconsultations/DH_4135308.
17. http://www.cochrane.org/.
18. http://www.jr2.ox.ac.uk/bandolier/.
19. http://www3.interscience.wiley.com/cgi-bin/mrwhome/106568753/HOME.
20. http://www.crd.york.ac.uk/crdweb/Home.aspx?DB=DARE&SessionID=&SearchID=&E=0&D=0&H=0&SearchFor=.

21. http://www.york.ac.uk/inst/crd/htadbase.htm.
22. http://www.crd.york.ac.uk/crdweb/Home.aspx?DB=NHS%20EED&SessionID=&SearchID=&E=0&D=0&H=0&SearchFor=.
23. http://www.library.nhs.uk/duets.
24. http://www.library.nhs.uk/specialistlibraries/.
25. http://www.library.nhs.uk/cancer.
26. http://www.library.nhs.uk/cardiovascular.
27. http://www.library.nhs.uk/childHealth.
28. http://www.library.nhs.uk/commissioning.
29. http://www.library.nhs.uk/guidelinesFinder/.
30. http://guidance.nice.org.uk/.
31. http://www.guideline.gov/.
32. http://www.nzgg.org.nz/.
33. http://www.nhmrc.gov.au/.
34. http://cks.library.nhs.uk/home.
35. http://www.nih.gov/.
36. http://www.pubmedcentral.nih.gov/.
37. http://ukpmc.ac.uk/ppmc-localhtml/news.html#obama.
38. http://publicaccess.nih.gov/policy.htm.
39. http://ukpmc.ac.uk/.
40. http://www.mrc.ac.uk/.
41. http://www.library.nhs.uk/booksandjournals.
42. http://www.apho.org.uk/.
43. http://www.nepho.org.uk/.
44. http://www.library.nhs.uk/PUBLICHEALTH/.
45. http://www.library.nhs.uk/.
46. http://www.nice.org.uk/guidance/index.jsp.
47. http://cks.library.nhs.uk/.
48. http://www.nphl.nhs.uk/.
49. http://www.comscore.com/.
50. http://www.nielsen-online.com/.
51. http://www.hitwise.com/.
52. http://www.silicon.com/search/?q=phorm&c=All&ss=date&submit.x=0&submit.y=0.
53. http://news.bbc.co.uk/2/hi/technology/7979378.stm.
54. http://www.indymedia.org.uk/.
55. http://www.browsermedia.co.uk/2009/02/26/is-social-media-taking-over-our-lives/.
56. http://medicalstudentblog.co.uk/twitter-doctors-medical-students-and-medicine-related/.
57. See Open Platform launch presentation shown at the Open Platform launch event at Kings Place in London on Tuesday, 10 Mar 2009.
58. http://www.guardian.co.uk/data-store.
59. http://www.nhs.uk/News/Pages/NewsIndex.aspx.
60. http://healthmap.org/en.
61. Brownstein JSJ, Clark C, Freifeld BS, Madoff LC. Digital disease detection—harnessing the web for public health surveillance. NEJM. 2009;360(21):2153–7. http://content.nejm.org/cgi/content/full/360/21/2153
62. http://www.statistics.gov.uk/default.asp.
63. http://www.statisticsauthority.gov.uk/.
64. http://www.ons.gov.uk/about-statistics/ns-standard/roles/gss/index.html.
65. http://www.statistics.gov.uk/hub/health-social-care/index.html.
66. http://www.hpa.org.uk/webw/HPAweb&Page&HPAwebAutoListName/Page/1201767910606?p=1201767910606.
67. http://www.pubmedcentral.nih.gov/articlerender.fcgi?tool=pubmed&pubmedid=18612462.

68. GE Healthcare to include urgent public health alerts in electronic medical records. http://money.aol.com/news/articles/qp/pr/_a/ge-healthcare-to-include-urgent-public/rfid201250137.
69. http://en.wikipedia.org/wiki/Collective_intelligence#cite_ref-15.
70. Flew, T. New media: an introduction. 3rd ed. Melbourne: Oxford University Press; 2008.
71. Zadeh LA. Fuzzy logic, neural networks, and soft computing. Comm. ACM-AI 1994;37: 77–84.
72. http://www-bisc.cs.berkeley.edu/zadeh/papers/ANoteonWebIntelligence,WorldKnowledgeand-FuzzyLogic-2004.pdf.
73. http://www.nature.com/nature/journal/v457/n7232/pdf/nature07634.pdf.
74. http://www.google.org/flutrends/.
75. Arjona J, Corchuelo R, Pena J, Ruiz D. Coping with web knowledge. In: Advances in web intelligence. Berlin: Springer; 2003. p. 165–78.
76. http://www.w3.org/2001/sw/.
77. http://www.opencyc.org/.
78. http://www.w3.org/TR/owl-guide/.
79. http://www.openclinical.org/ontologies.html.
80. Bishop CM. Pattern Recognition and Machine Learning. Springer; 2007.
81. http://www.autonomy.com/.
82. http://en.wikipedia.org/wiki/Bayesian_inference.
83. http://en.wikipedia.org/wiki/Information_theory.
84. Novak V, Perfilieva I, editors. Discovering the world with fuzzy logic. Studies in fuzziness and soft computing. Heidelberg: Physica-Verlag; 2000.
85. Zadeh LA. From computing with numbers to computing with words from manipulation of measurements to manipulation of perceptions. IEEE Transactions on Circuits and Systems 1999;45(1):105–19.
86. Bargiela A, Pedrycz W. Granular Computing-An Introduction. Boston, Dordrecht, London: Kluwer Academic Publishers; 2003.
87. http://www.bigpedia.com/encyclopedia/Lotfi_Zadeh.
88. http://sipi.usc.edu/~kosko/.
89. http://modo.ugr.es/en/soft_computing.
90. http://www.google.com/ads/preferences.
91. http://www.informationprescription.info/index.html.
92. http://talk.nhs.uk/.
93. http://www.direct.gov.uk/.
94. http://www.medpedia.com/.
95. https://www.healthspace.nhs.uk/visitor/visitor_healthandlifestyle.aspx.
96. http://www.healthvault.com/.
97. https://www.google.com/health.
98. http://www.boston.com/news/nation/washington/articles/2009/04/13/electronic_health_records_raise_doubt/.
99. http://www.dh.gov.uk/PublicationsAndStatistics/Publications/PublicationsPolicyAndGuiance/PublicationsPolicyAndGuidanceArticle/fs/en?CONTENT_ID=4094550&chk=aN5Cor.
100. http://www.foresight.gov.uk/OurWork/ActiveProjects/Obesity/KeyInfo/Index.asp.
101. http://www.noo.org.uk/.
102. http://www.sepho.org.uk/.
103. Cole TJ, Freeman JV, Preece MA. Body mass index reference curves for the UK, 1990. Archives of Disease in Childhood 1995;73:25–29.
104. http://www.library.nhs.uk/PUBLICHEALTH/ViewResource.aspx?resID=304971.
105. http://www.library.nhs.uk/PUBLICHEALTH/ViewResource.aspx?resID=300242.
106. http://www.library.nhs.uk/PUBLICHEALTH/ViewResource.aspx?resID=311440.
107. http://www.library.nhs.uk/PUBLICHEALTH/ViewResource.aspx?resID=311497.
108. http://www.library.nhs.uk/PUBLICHEALTH/ViewResource.aspx?resID=311490.
109. http://www.library.nhs.uk/PUBLICHEALTH/ViewResource.aspx?resID=311398.

110. http://www.library.nhs.uk/PUBLICHEALTH/ViewResource.aspx?resID=311534.
111. http://www.library.nhs.uk/PUBLICHEALTH/ViewResource.aspx?resID=311403.
112. http://www.library.nhs.uk/PUBLICHEALTH/ViewResource.aspx?resID=311400.
113. http://www.library.nhs.uk/PUBLICHEALTH/ViewResource.aspx?resID=311522.
114. http://www.library.nhs.uk/PUBLICHEALTH/ViewResource.aspx?resID=311422.
115. http://www.library.nhs.uk/PUBLICHEALTH/ViewResource.aspx?resID=311426.
116. http://www.library.nhs.uk/PUBLICHEALTH/ViewResource.aspx?resID=311371.
117. http://www.library.nhs.uk/PUBLICHEALTH/ViewResource.aspx?resID=301004.
118. http://www.childgrowthfoundation.org/default.htm.
119. http://www.nationalobesityforum.org.uk/.
120. http://www.opsi.gov.uk/acts/acts2008/ukpga_20080014_en_1.
121. http://nationalobesityforum.org.uk/children-mainmenu-176/395-an-approach-to-primary-prevention-of-obesity-in-children-and-adolescents.html.
122. Data Retention (EC Directive) Regulations 2009. http://www.opsi.gov.uk/si/si2009/draft/ukdsi_9780111473894_en_1.
123. Communications Data Bill. http://www.commonsleader.gov.uk/output/page2667.asp.
124. http://www.cabinetoffice.gov.uk/cio/transformational_government/open_source.aspx.
125. [http://www.dh.gov.uk/PublicationsAndStatistics/Publications/PublicationsPolicyAndGuiance/PublicationsPolicyAndGuidanceArticle/fs/en?CONTENT_ID=4094550&chk=aN5Cor.

Healthcare Prosumerism

A. C. M. Dumay and J. L. T. Blank

Abstract A vision is presented to relief the healthcare system of the top of its burden by bringing healthcare closer to the customer—to bring healthcare into the home—with healthcare prosumers rather than consumers, priority medical devices rather than scarcity of labour and market incentives to create fair prices; Social and ethical considerations to deal with affordability, accessibility and efficacy of healthcare at home. The chapter concludes with a high-level research agenda for biomedicine and bioinformatics.

1 Introduction

The demand for healthcare under chronic conditions increases rapidly in volume and form. Due to ageing and growth of the population in the Netherlands [13], the number of persons with a chronic disease will increase in the next 20 years, with the largest increase expected in the number of persons with diabetes and osteoporosis. In the next 20 years, 300,000 additional cases of diabetes are expected, with a further 100,000 cases of diabetes expected if the increase in the prevalence of obesity continues at the present rate. 350,000 additional cases of osteoporosis are expected in the next 20 years. In the past decades, smoking has

A. C. M. Dumay (✉)
TNO Quality of Life, Delft, The Netherlands
e-mail: A.C.M.Dumaij@tudelft.nl

A. C. M. Dumay and J. L. T. Blank
Innovation and Public Sector Efficiency Studies, Delft University of Technology, Delft, The Netherlands

Commun Med Care Compunetics (2011) 1: 43–52
DOI: 10.1007/8754_2010_3

Published Online: 12 November 2010

declined in men, while it has increased in women, resulting in future patterns of smoking-related diseases (COPD and lung cancer) that are different for men and women: a smaller increase in men, and a larger increase in women than expected on the basis of demographic changes alone. These are some of the results drawn from projections of the chronic disease burden for 2005 up to 2025. Projections—for cardiovascular diseases (myocardial infarction, stroke and heart failure), diabetes mellitus, cancer (lung, colon, breast), asthma, COPD and osteoporosis—were made for anticipating future healthcare needs and costs [5].

The supply of high quality care is already hampered by lack of qualified practitioners and will face a serious shortage in the next decades. Scientific advancement brings about new diagnostic methods and technologies and consequently further increases healthcare demand. The cost of healthcare, medical technology and medicines increases to a critical portion of the gross national product of nations. The efficiency growth in the healthcare delivery of organisations and the healthcare sector has come to a standstill since the healthcare system cannot adopt innovations at large (i.e. national) scale.

In brief: As of year 2010 over one million people are employed in healthcare sector in the Netherlands while in 2025 there will be no professional healthcare available of 55 out of 145 somatic patients. There will be a shortage of 40% of labour [17]. Healthcare providers are expected to increase efficiency by at least 1.5% per year from 1995 but fall short with respect to this target [25].

In our vision the top of this productivity gap can be relieved by bringing healthcare closer to the customer—to bring healthcare to the home—with healthcare prosumers rather than consumers who apply self management.

Questions to elaborate on are "Does the healthcare consumer want prosumerism?", "What is the role of technology?", "What are mechanisms to form new prices?", "What are social and ethical considerations to take into account?", "What market structure and governance model does right to all parties involved?"

In this chapter a vision is presented, evaluated and made concrete with a strategic research agenda. Section 2 postulates the vision and Sect. 3 describes the role of technology in that vision. Considerations from the point of view of health economy, society and ethics, and governance are presented in Sects. 4, 5 and 6, respectively. The chapter concludes with the strategic research agenda.

2 Vision: Prosumerism and Self Management

In recent years a new societal requirement has appeared: the requirement for support. People want support to structure and organize their lives such that they live in their own fashion. To satisfy that requirement it is necessary to personalize services and support each person in a personalized way. Manufacturing and services more and more include the end user, not only in the design process but in operations as well: bread is baked in your own kitchen, bank accounts are managed from your home, etc. A principle known as prosumerism. This phenomenon took flight with the rise of the internet because information and communication technologies influence service delivery and the point of contact with the consumer [41].

The healthcare system is not designed to meet this requirement because the patient is not part of the system, rather a subject. The new requirement can only be satisfied with a new way of work. This is what Zuboff and Maxmin call a Copernican transition: the consumer with his or her needs comes from the periphery into the centre of the system, while suppliers with their products and services migrate from the centre to the periphery. In other words: a demand-driven supply following the needs of the consumer.

In healthcare prosumerism is synonymous with an active form of self management (in the literature also referred to as self care). Self management was introduced by Lau who defines it as stimulating the responsibility of the patient with the aim to maximise his or her potential to health and well being [20]. Self management is a new dimension to the concept of patient empowerment in which a patient co-operates with his or her healthcare practitioner to improve health outcomes [3]. It supports the Copernican transition with first evaluation studies [22]. However, self management can only be performed when the patient has a sufficient understanding of his or her health conditions [33]: What kind of disease? What are the consequences and prospective? What can the patient do and should not do? How to assess physiological functional state? Which guidelines to apply and when to seek help? What to eat? How much exercise today? Which medicine at what time to take? (Halme et al 2005). Also, informed decision support [32] or coaching is a critical aspect of self management, regardless of the fact that the support comes from a peer, health companion, practitioner of (medical) device [29]. Coaching by a health practitioner appeared to be effective when it follows the stages of the disease [21, 38]. The focus is than not on applying medical interventions but on behavioural interventions [10].

Self management itself is a dynamic process and it affects the utilization of services, patient satisfaction, and health outcomes. Having a valid and reliable measure is important to understanding the effects of self management. However, there is still a lack of a standardized measurement instrument for the empirical assessment of self management in a general patient population [24]. A number of recent initiatives are taking place to develop self management measurement instrument for the general population. Loukanova and Bridges [23], e.g., have proposed a self management instrument for the general population covering six domains such as knowledge, access, advocacy, decision making, health status and outcomes, and literacy.

Recent studies reveal that most somatic patients chose for self management with support of a health practitioner who is co-operative, empathic and communicative [6]. Operational guidelines for practitioners are available [39] based on motivational interviewing techniques [27].

3 The Role of Technology

Technology in healthcare is often referred to as medical device. For the legislators of the European Commission *a 'medical device' means any instrument, apparatus, appliance, material or other article, whether used alone or in combination, including the software necessary for its proper application intended by the manufacturer to be used for human beings for the purpose of:*

- *diagnosis, prevention, monitoring, treatment or alleviation of disease;*
- *diagnosis, monitoring, treatment, alleviation of or compensation for an injury or handicap;*
- *investigation, replacement or modification of the anatomy or of a physiological process;*
- *control of conception,*

and which does not achieve its principal intended action in or on the human body by pharmacological. immunological or metabolic means, but which may be assisted in its function by such means. 'Accessory' means an article which whilst not being a device is intended specifically by its manufacturer to be used together with a device to enable it to be used in accordance with the use of the device intended by the manufacturer of the device [11].

This medical device directive (MDD) differentiates technology into various groups and risk classes by its intended use. Custom made device means any device specifically made in accordance with a duly qualified medical practitioner's written prescription which gives, under his responsibility, specific design characteristics and is intended for the sole use of a particular patient. Mass-produced devices which need to be adapted to meet the specific requirements of the medical practitioner or any other professional user are not considered to be custom-made devices. Device intended for clinical investigation means any device intended for use by a duly qualified medical practitioner when conducting investigations in an adequate human clinical environment. For the purpose of conducting clinical investigation, any other person who, by virtue of his professional qualifications, is authorized to carry out such investigation shall be accepted as equivalent to a duly qualified medical practitioner [11]. The obligations of this directive to be met by manufacturers also apply to the natural or legal person who assembles, packages, processes, fully refurbishes and/or labels one or more ready-made products and/or assigns to them their intended purpose as a device with a view to being placed on the market under his own name. It is up to the competent authority (in the Netherlands: Dutch Inspectorate for Healthcare IGZ) to decide whether or not the MDD is applicable to a specific product. If so, product standards, production standards and service directives apply depending on the risk class of the product. These are drawn from the public domain as the European Committee for Standardization (CEN), the International Organization for Standardization (ISO), the American National Standards Institute (ANSI), the many standardization committees of the Institute of Electrical and Electronics Engineers (IEEE), Health Level 7 (HL7), etc., and national standardization organizations. Only a notified body is authorized to test the product against applicable standards and provide certification marks. A post-market surveillance system reports incidents and accidents with specific devices world wide, e.g., by the ECRI Institute (www.ecri.org). There are already over-the-counter medical devices that do not comply to product standards [18, 40] and will affect health conditions in the long run.

Today, the process of bringing hospital equipment into the home is in full swing. Examples are glucose measurement by blood glucose monitor (BGM) for diabetes patients, ventilators for COPD patients, ECG-monitoring for CHF patients, etc. These medical technologies are often combined with telemetry with a practitioner for monitoring purposes. This so-called eHealth (also referred to as telemedicine or telemonitoring) plays an important role in selfcare [2]. eHealth application in case of comorbidity is still in its infancy. Another group of technology to support selfcare is ambient assisted living technology with smart homes as a result. Demiris and Hensel define "a 'smart home' is a residence wired with technology features that monitor the well-being and activities of their residents to improve overall quality of life, increase independence and prevent emergencies." [14]. Smart homes offer a promising and cost-effective way of improving home care for older adults in a non-obtrusive way, which allows greater independence, maintaining good health and preventing social isolation. Smart homes are equipped with sensors, actuators, and/or biomedical monitors. These devices operate in a network connected to a remote centre for data collection and processing [8]. Accordingly, Smart homes become more and more popular and receive more and more attention as a support environments for healthy, socially participating and self-caring inhabitants [1, 4, 26].

To empirically establish the effect of health technologies on clinical outcomes, randomized controlled trials (RCT) are considered the "gold standard". However, various RCTs face the Eysenbach Law of Attrition, i.e., the phenomenon of participants drop out of the study [16]. There are various reasons why people may drop out of a study, e.g., moving house, rapid change of health conditions (either positive of negative), death, loss of interest and motivation, social pressure, inconvenience, etc. [12]. The Law of Attrition especially applies when conducting RCTs of technology supporting self-care, since two related tests are performed: one on self care and one on the technology supporting the self care. Researchers cannot benefit from laboratory experiments where people are subject to controlled tests during a limited time. Instead they create an experimental setting that elicits representative use of technology supporting self care "in the wild", i.e. at home, over a prolonged period of time, and based on intrinsic motivation (Deci and Ryan 2005) to elicit natural and longitudinal effects of technology. Blanson Henkemans et al. [4] found that intrinsic motivation is a critical inclusion factor to prevent the Law of Attrition, knowing that a bias is introduced in the test results. Most of the evaluation studies are pilot or short-term projects, consisting of nonrandomized trials without control groups, which often show methodological weaknesses (e.g., in samples size, context, and study design), limiting the generalization of the findings [31].

4 Economic Considerations

Aside from a technical psychological and societal perspective the issue of self management and prosumerism can also be considered from a pure economical point of view. In a conceptual article on the production of public services Ostrom [30] indicates public sector users as co-producers, referring to the fact that education, health care and law enforcement can only be effective if users are willing to put some serious effort in it as well.

Since prosumerism can be regarded as a form of resource substitution and technological change, the economic aspects fit very well in the theory of production. In this theory technology reflects the technical transformation from resources into services. In a neoclassical framework the (optimal) allocation of resources and services depends on the existing technology, services prices and resource prices. According to this theory increasing resource prices, for instance of nursing personnel, may lead to an increasing usage of equipment or input of co-producers. However, there are a few striking differences between the usage of "normal" resources and co-producers.

There is no market and no market price for co-producers which may result in an over usage of co-producers, indicated as allocative inefficiency. Health care providers are inclined moving into the direction of "free" co-producers' usage. Since it is obvious that co-producers also value their time and efforts this may lead to welfare losses.

In particular, co-producers' resource price for curative care patients may be substantial due to relative high instruction time. Co-producing for chronically disease patients may thus be more suitable. Sunk cost for co-producers can be spread over a longer period of time resulting in lower co-producers' resource prices.

Another difference refers to the phenomenon of economies of scale. Co-producing mostly takes place outside the hospital or medical care facility. Consequently, medical devices and equipment may be left unused a substantial period of time during the day. So substantial scale inefficiencies may prevail.

The heterogeneity of quality of co-producers is probably large resulting in technical efficiencies. Errors in medical devices or medical drug usage put an extra burden on medical care, for instance due to re-admissions.

It is obvious that aside from the consequences for patients, the technology of health service delivery goes through major changes and also affects the way doctors and nursing personnel are operating and the way medical processes are managed and controlled.

Prosumerism is often presented as a Columbus' egg for the problems of extensive growth in health care expenditure and labour market shortages. However, this is debatable. From an economic reasoning this type of technology change and resource substitution may lead to substantial allocative, scale and technical inefficiencies. Substantial welfare losses can be expected. The risk of the prosumerism strategy is that these inefficiencies may be unrevealed, whereas efficiency gains for the health care providers may seem straightforward. Social benefits probably only occur in the segment of chronically diseases with treatments including low cost medical devices, simple organizational changes and low cost training of patients.

5 Social and Ethical Considerations

Successful self-management can avert preventable mortality and mortality, improve the quality of life of individuals and families, and boost economic productivity. It is recognized as a core management process that is effective in improving the quality of care, especially in combination with other processes, such as case management [7].

A critical consideration to take into account is "preparedness" of the patients. Studies have revealed that a up to a third of all patients in the USA with a chronic disease such as diabetes mellitus, asthma and hypertension have not received effective therapy, including self management skills [37]. Also, 42% of persons with diabetes have never been advised or are still confused about how to manage their disease; only 50–80% of people with type II diabetes achieve optimal glycemic control, and only 50% of patients with chronic disease provide a self-management programme [28]. Whose responsibility is it to assure that patient autonomy to self-manage is supported, the harms of poor self-management are avoided, and the patient is well-prepared to start self-managing? What is the ethical framework for these situations, with ethics referring to standards of behavior that tell us how human beings ought to act in the many situations in which they find themselves-as friends, parents, children, citizens, businesspeople, teachers, professionals?

Classical healthcare is delivered within a frame of trust between the healthcare practitioner and the patient. There are complementary roles within that relationship. The healthcare practitioner brings knowledge, experience and judgment to diagnose of intervene the disease; the patient brings knowledge and experience about the reactions of his or her body to the treatment and how he or she (intends to) integrates the disease in daily activities and job functions. Redman [34] argues that an ethical structure and a set of goals and expectations for these complementary roles are essentially based on principles of autonomy, doing no harm (*primum non nocere* of the Hippocratic oath) and allocation of resources. She also advocates an increasing role of nurses in providing self-management preparation and as such share the responsibility with other healthcare professionals. The same questions about preparedness apply when prosumerism is advocated.

6 Governance

The final aspect of healthcare prosumerism or self management is the question "Who is responsible and accountable for what?" Today, when a healthcare practitioner accepts a patient for treatment he/she is fully accountable for all actions taken, whether this is diagnosis, treatment or referral. The activities are covered by a medical treatment agreement between practitioner and patient. If no such agreement exists explicitly, the Dutch law still assumes a default agreement based on informed consent (Nieuwenhuis et al. 2005). See also The Health Insurance Portability and Accountability Act (HIPAA) of 1996. Hospitals, nursing homes and home care services are constructs based on this treatment agreement. If a patient does not comply to the doctor's prescriptions, the treatment is hampered, slowed down or even disrupted. The patient bares the consequences. In case a patients buys healthcare over the counter without legal consultation of a healthcare practitioner or mixes treatment by a practitioner with healthcare over the counter a whole new situation arises for which new governance structures are required. This is one of the main reasons why adoption by healthcare practitioners of self management concepts is slow [35]. Safety of consumer products must be guaranteed, supply of necessities and spare parts of the products must be managed, privacy guaranteed, accountability and supervision must be in place, quality ratings must be transparent.

7 A Strategic Research Agenda for Biomedicine and Bioinformatics

Prosumerism requires transparency of objectives, means and information and a transition at the level of the healthcare system. In the field of biomedicine and bioinformatics the focus should be on personalised diagnosis and interventions [9], i.e. precision diagnosis and precision intervention rather than workshop trial and error interventions. Personalization stimulates the patient to take an active role in the healthcare process.

New classes of sensor suites must be developed to extract biomedical and biochemical data from the patient for each chronic disease. The data should support medication, technical and life style interventions to improve monitoring, reduce visits to a clinic and prevent hospitalization. Parameters must of course be clinically relevant and measured with the right precision and accuracy. On the other hand, new parameters stimulate biomedical research to derive new interventions.

Information exchange between patient and practitioner and information exchange among practitioners must be made transparent, eligible and timely [15]. Standardized consumer health vocabularies must be available to advance patient care through the delivery of a dynamic and sustainable, scientifically validated terminology and infrastructure that enables clinicians, researchers and patients to share health care knowledge worldwide, across clinical specialties and sites of care. Systematized Nomenclature of Medicine-Clinical Terms (SNOMED CT) is considered to be the most comprehensive, multilingual clinical healthcare terminology in the world [19].

Further research is required on the absorption potential and dissemination strategies of prosumerism [36].

References

1. Ackerman MJ. The smart home. J Med Pract Manag. 2009;25:68–9.
2. Alpay LL, Blanson Henkemans O, Otten W, Rövekamp AJM, Dumay ACM. eHealth applications for patient empowerment: current state, trends and directions for future developments. J Telemed Telecare. 2010. (Accepted).
3. Aujoulat I, Marcolongo R, Bonadiman L, Deccache A. Reconsidering patient empowerment in chronic illness: a critique of models of self-efficacy and bodily control. Soc Sci Med. 2008;66(5):1228–39.
4. Blanson Henkemans OA, Alpay LL, Dumay ACM. Aging in place: self-care in smart home environments. In: Al-Qutayri MA, editor. Smart home systems. Olajnica: In-Tech; 2010. p. 105–20. ISBN 978-953-307-050-6.
5. Blokstra A, Verschuren WMM. Vergrijzing en toekomstige ziektelast. Prognose chronische ziektenprevalentie 2005–2025. Report RIVM 260401004/2007. 2007. http://www.rivm.nl/bibliotheek/rapporten/260401004.html.
6. Carmel M, Peterson C, Robinson R, Sturmberg JP. Care for chronic illness in Australian general practice–focus groups of chronic disease self-help groups over 10 years; implications for chronic care systems reforms. Asia Pacific Fam Med. 2009;8:1.doi:10.1186/1447-056X-8-1.
7. Casalino L, Gillies RR, Shortell SM, Schmittdiel JA, Bodenheimer T, Robinson JC, Rundall T, Oswald N, Schauffler H, Wang MC. External incentives, information technology and organized processes to improve health care quality for patients with chronic diseases. JAMA. 2003:289:434–41.
8. Chan M, Campo E, Esteve D, Fourniols J-Y. Smart homes—current features and future perspectives. Maturitas. 2009;64:90–7.
9. Christensen CM, Grossman JH, Hwang J. The innovator's prescription. A disruptive solution for health care. New York: McGraw-Hill; 2009. ISBN 0-07-159208.
10. Corbin J, Strauss A. Unending work and care: managing chronic illness at home. San Fransico: Jossey-Bass Publishers; 1988. ISBN 1555420826.
11. Council Directive 93/42/EEC of 14 June 1993 concerning medical devices. http://europa.eu/legislation_summaries/internal_market/single_market_for_goods/technical_harmonisation/l21010b_en.htm.
12. Davis M, Addis M. Predictors of attrition from behavioral medicine treatments. Ann Behav Med. 1999;21:339–49.
13. De Bakker DH, Polder JJ, Sluijs EM, Treurniet HF, Hoeymans N, Hingstman L, Poos MJJC, Gijsen R, Griffioen DJ, van der Velden LFJ. Op een lijn—Toekomstverkenning eerstelijnszorg 2020 (Public health forecast for primary care in the Netherlands in 2020). 2007. http://www.rivm.nl/bibliotheek/rapporten/270751009.pdf.
14. Demiris G, Hensel BK. Technologies for an aging society: a systematic review of "smart home" applications. In: Yearbook of medical informatics. 2008. p 33–40.
15. Dumay ACM, Freriks G. Towards developing a coherent healthcare information infrastructure. In: E-Health in Belgium and in the Netherlands, Proceedings of MIC2002, Studies in Health Information and Informatics, 93. Amsterdam: IOS Press; 2002. p. 1–7. ISBN 0926 9630.
16. Eysenbach G. The law of attrition. J Med Internet Res. 2005;7(1):e11.
17. Goris A, Mutsaers H. Ruimte voor arbeidsbesparende technologie om in 2025 voldoende zorg te bieden. Leiden: STG/Health Management Forum; 2008. ISBN 978-90-75425-17-8.
18. Hollestelle ML, Hilbers ESM, van Drongelen AW. Risks associated with the lay use of 'over-the-counter' medical devices. Study on infrared thermometers and wound care

products. RIVM BMT letter report 360050002. 2007. http://www.rivm.nl/bibliotheek/rapporten/360050002.pdf.
19. IHTSDO. International Health Terminology Standards Development Organization. 2010. http://www.ihtsdo.org/.
20. Lau DH. Patient empowerment: a patient-centred approach to improve care. Hong Kong Med J. 2002;8(5):372–4.
21. Leventhal H, Weinman J, Leventhal EA, Phillips LA. Health psychology: the search for pathways between behavior and health. Ann Rev Psychol. 2008; 59:477–505.
22. Lorig K, Holman M. Self-management education: history, definitions, outcomes and mechanisms. Ann Behav Med. 2003;26:1–7.
23. Loukanova S, Bridges JFP. Measuring patient empowerment in the general community. Paper presented at the annual meeting of the Economics of Population Health: Inaugural Conference of the American Society of Health Economists, TBA. Madison; 4 June 2006.
24. Loukanova S, Bridges JFP. Empowerment in medicine: an analysis of publications trends 1980–2005. Cent Eur J Med. 2008;3(1):105-10.
25. Luijks K, Putters K, de Roo A. Verhogen van de arbeidsproductiviteit in de zorgsector. Verkenning van de mogelijkheden en beperkingen. Tilburg: Tranzo; 2005. ISBN 9072725999.
26. Martin S, Kelly G, Kernohan WG, McCreight B, Nugent C. Smart home technologies for health and social care support. Cochrane database of systematic reviews. 2008. http://www.cochrane.org/reviews/en/ab006412.html.
27. Mesters I. Motivational interviewing: hype or hope. Health Expect. 2009;5:3–6.
28. Norris SL, Nichols PJ, Caspersen CJ, Glasgow RE, Engelgau MM, Jack L, Snyder SR, Carande-Kulis VG, Isham G, Garfield S, Briss P, McCulloch D. Increasing diabetes self-management education in community settings; a systematic review. Am J Prev Med. 2002;22(4S):39–66.
29. O'Connor AM, Bennett CL, Stacey D, Barry M, Col NF, Eden KB, Entwistle VA, Fiset V, Holmes-Rovner M, Khangura S, Llewellyn-Thomas H, Rovner D. Decision aids for people facing health treatment or screening decisions. Cochrane Database of Systematic Reviews 2009, Issue 3. Art. No.: CD001431. 2009. doi:10.1002/14651858.CD001431.pub2.
30. Ostrom E. Institutional arrangements and the measurement of police consequences in urban areas. Urban Affairs Quart. 1971;(6):447–75.
31. Pare G, Jaana M, Sicotte C. Systematic review of home telemonitoring for chronic diseases: the evidence base. JAMIA. 2007;14:269.
32. Raats CJ, van Veenendaal H, Versluijs MM, Burgers JS. A generic tool for development of decision aids based on clinical practice guidelines. Patient Educ Couns. 2008;73:413–7.
33. Redman BK. Patient self-management of chronic disease. Sudbury, MA: Jones and Barlett; 2004.
34. Redman BK. The ethics of self-management preparation for chronic illness. Nursing Ethics. 2005;12(4);360–9.
35. Reti S, Feldman H, Safran C. Governance for personal health record. JAMIA. 2008;16:14–7.
36. Rogers EM. Diffusion of innovations. New York: Free Press; 2003. ISBN 978-0-7432-2209-9.
37. Rothman AA, Wagner EH. Chronic illness management: what is the role of primary care? Ann Intern Med. 2003;138:256–61.
38. Ryan RM, Deci EL. Self-determination theory and the facilitation of intrinsic motivation, social development, and well-being. Am Psychol. 2000;55(1):68–78.
39. Schaefer J, Miller D, Goldstein M, Simmons L. Partnering in self-management support: a toolkit for clinicians. Cambridge: Institute for Healthcare Improvement; 2009. http://www.improvingchroniccare.org/downloads/partnering_in_selfmanagement_support__a_toolkit_for_clinicians.doc.
40. Slingerland RJ, Miedema K. Evaluation of portable blood glucose meters. Problems and recommendations. Clin Chem Lab Med. 2003;41(9):1220–3.
41. Zuboff S, Maxmin J. The support economy—why corporations are failing individuals and the next episode of capitalism. New York: Penguin; 2004.

Patient Expectations in the Digital World

Lodewijk Bos

Abstract This chapter is based on my keynote during the Estonian e-Health Conference in Tallinn, october 2010. The presentation was given wearing two hats, one as president of ICMCC, the other as a patient.

Keywords Compunetics · Patient · Narrative · Electronic Health Records · Physician-patient Relationship

1 Historic Overview of ICMCC

In 2004 I became founding president of ICMCC, the International Council on Medical and Care Compunetics; a foundation dealing with the social, societal and ethical aspects of the use of computing and networking. (compunetics) in medicine and care.

Our initial focus was on two points: awareness and the supply of validated information, targeting both patients and professionals. Soon we added two more specific items, electronic records and digital homecare.

In 2006 we created a first kind of knowledge center on the internet, our Record Access Portal [1], followed by our guidelines on Patient Record Access to the WHO [2].

Early 2007 we started the ICMCC News Page, which since has become one of the leading sources on the internet for news on health information technology.

L. Bos (✉)
ICMCC, Utrecht, The Netherlands
e-mail: lobos@icmcc.org

Commun Med Care Compunetics (2011) 1: 53–59
DOI: 10.1007/8754_2010_12

Published Online: 11 January 2011

Last year we added to it the most comprehensively linked scientific article database. The ICMCC website is estimated to reach an audience of about 134,000 unique visitors in the year 2010.

We have our highly regarded proceedings book series with IOS Press and our own series "Communications in Medical and Care Compunetics" with Springer Verlag.

2 Perception

The title of this presentation, patient expectations in the digital world, seems to indicate that patient expectations are, or at least could be different in a digital world. To be honest, I'm not so sure.

In march 2006, I needed to have surgery. I went into the operation theater diagnosed with a hemorrhoid. That diagnosis changed during the following hours into an extremely aggressive kind of Non-Hodgkin Lymphoma, as I learned a week later.

That is the moment when you discover that with all developments in the digital world, there is one thing that will never change; you want to be cared for, if possible even cured. And it is funny to see, looking back, how the digital world seems to have confused things on both sides of the treatment table.

A week after the surgery, I was told by my surgeon it was cancer. I asked for the pathology report which gave as sole indication: malignant. The next morning I was seen by the oncologist. After 3 min I asked him not to speak his usual patient lingo. So he immediately came to the conclusion that I was one of those "knowledgeable" patients, you would call them e-patient now. Consequently he asked: then I suppose that you have already looked things up on the internet? I looked at him and told him that my copy of the pathology report only indicated malignant, hardly worth a google search. He then informed me that it was a very rare and aggressive type of Non-Hodgkin. Of course I immediately asked for his version from pathology.

I must admit that I did not grasp what it meant he said, that dawned on me much later. But having gone through my share of medical professionals in my life, I asked if he knew what he was talking about. To avoid being entirely misunderstood i added that he had only seen the pathology report but not the tumor previous to surgery since it was not recognized as such. So I told him that I could illuminate him and opened my laptop to show him a picture of the supposed hemorrhoid made a couple of hours before the operation. Weeks later he admitted that due to the shock caused by that picture he had forgotten to perform the standard body check and to order the proper blood analysis.

So he had misunderstood the concept of being an informed patient and I had misjudged the power of images.

3 Information

In my youth it was a common saying that a disease is already half cured when it is given its proper name. And even in the google era that is still true. When a friend of mine was told she had breast cancer, she immediately said to me that she was going to search on the internet. I asked if she had been told the details of her cancer. She hadn't, but was going to look anyway. A couple of hours later she came back to me and said, you were right, its only misery I find or useless information.

So the old adagio still holds, you have to have the basic information to make yourself knowledgeable.

So the principles of care and information have not changed. What did change, however, are the availability of information and the methods of care. What used to be a long trip to a good, scientific library is now only a mouse click away. But you still have to know where to look.

In the past your social network, your family, friends, neighbors, taught you the small household tricks for minor ailments, from alcohol to numb you for a small surgical interaction to chicken soup for the flu. And you knew that the journals you read at the barbershop did not give you good and valid medical and health information, if any.

Your local doctor was the man who supposedly knew almost all. That in many cases he only stimulated your self-healing capacity and gave you a blue or a pink sugar pill you either did not realize or you did not want to know.

Till two generations ago there were hardly any car or plane crashes. Alzheimer's disease, obesity, cancer were relatively rare. People did not grow as old as they do now; the standard for quality of life and the way of living were much different; one had to move to get somewhere and to work to keep warm. And people were not hindered by knowledge. We were not able to measure the way we do now and being overweight was considered a sign of prosperity in many countries and cultures. People followed the tradition of experience learned from generations before and trusted the wisdom of one of the only three educated people in their environment, the doctor (the other two being the priest and the teacher).

Modern communication technology, the internet, has brought us tools to gather data, has brought us access to information and ways to share experience far beyond what has been, by way of speaking, in our genes for hundreds if not thousands of years.

The last decade we have been seeing the development of a whole new societal concept, homo informaticus, the informed, communicating human.

We call it 2.0, which can be described as all things linked to and linked through active (i.e. social) communication.

Health also knows its 2.0 version. We at ICMCC have defined and published it in 2008 as the combination of health data and health information with (patient) experience through the use of ICT, enabling the citizen to become an active and responsible partner in his/her own health and care pathway [3].

As mentioned earlier, our societal concepts are changing. We receive far more information than ever before. Communication has changed to a completely different level. Our social structures are disappearing, being replaced by those enabled through communication, more specific the internet. Social networks like facebook replace the old neighborhood and enable sharing almost anything we want. Specific networks for patients are developing. The wisdom of generations is making way for the experiences from other patients all around the world even on the rarest of diseases.

Information has become easily accessible. Many people fear that the regular doctor will be replaced by Dr. Google, but recent surveys show that, until now, that is not the case [4]. However, patients go to the web to check on the information they got from their caregivers. This is the new version of the old adagio, when it has a name you can find information.

The problem is that most of that information is not validated and most of those who search for that information are not capable or trained to notice that lack of value. And since very often the originator of that information is carefully hidden, you don't know whether you are manipulated into certain actions. Many popular sites on various health problems are created to boost the industry. Sleeping problems, snoring, erectile dysfunction, all popular items which have become health problems without being properly diagnosed whilst cures are being offered by the pharmaceutical industry.

So the informed human turns into an informed patient, which is at the roots of what might become one of the most important paradigm shifts in health, the physician-patient relationship.

Patient centric has become the keyword for modern health. However, the patient is verbally put in that spot, but not in reality.

For health to become real patient centric—or participatory as it is being called these days—you have to change attitudes on both sides. For patients still will have to accept that doctors do have had years of training and practicing which gives them ample medical experience. On the other hand, caregivers have to accept that the patient is the one with the experience of being ill.

So each party brings in his or her own experience. Only when combining it with proper information it can become a workable, participatory relationship, as Dr. Hannan, a British GP called it, a partnership of trust [5].

The core of that shared information is the electronic health record.

4 Records

For years, watching the discussions rock on terminology, I have been careful in my choice of words. I said electronic health records not electronic medical records. And I did certainly not say personal health records. Despite all the discussions, the meaning is in the used terminology. Medical records contain the information gathered from medical procedures and medical opinions. That can be hospital exams, lab results, medication, surgery reports, a doctor's diagnosis.

Health records contain all health information, which means all information from the medical records combined with all other things that might be relevant to a holistic view of your health and well-being. And part of that is delivered by the patient or citizen himself. That is where the patient's experience comes in. For only the patient can describe what he or she is feeling, what the day-to-day problems are. And these observations should be combined with the proper medical information to make sense.

In my view the concept of personal records will not work for several reasons. Most of the medical data in a record is coming from sources to which a patient does not have direct access, or might not even be aware of. The idea behind a personal health record is that the patient is in charge of putting information into it, medical information included. There is, however, no control of that input, so it will never be known whether the information in such a record will be complete. A patient might, maybe even intentionally, have left out important information. Another reason is that it demands unnecessary human effort [6]. Every piece of medical information is stored in the place where it is created, hospital, laboratory, surgery, pharmacy. So why copy it somewhere else as well? In the UK it was recently discovered that records were updated at the places of origin, but not in the second, central storage, the so-called spine [7].

The legal storage period has always been limited, due to the fact that we were dealing with physical storage, paper, photographs, dental prints. Now that all that information is available in digital form, we can change the laws and make storage eternal. That way we can provide lifelong electronic health records at the same time creating the possibility of building a family history which is important considering the fact that genetic information will soon become a commodity.

5 Narrative

There remains of course the question how to handle the patient's narrative.

After I ended my chemotherapy I complained to my oncologist about rheumatic pains in my hands. I never suffered from rheumatism so why did I use that description? And how does my oncologist know what I mean? And how will that be put in my record in a meaningful way?

A human being is very often lazy and forgetful. How can we expect him or her to consistently record these so-called observations in daily living? That is another new aspect of the digital world. Within the next decade many of these will be mechanically recorded. Five years ago we were worrying about the intrusiveness of monitoring but now things are changing at an unbelievable speed.

So called smart technology will help to record necessary data, like blood pressure or glucose levels, but also remind us to take medication correctly and timely or register our physical movements, for example to prevent the elderly from falling. Smart phones, smart clothing will make it far less intrusive.

These technologies will be a big help for both patient and caregiver. For the latter it will mean saving time and resources, for the patient it will mean a major step forward towards self-management or independent living.

In this development there are a couple of hurdles to take. The data produced with these technologies will have to be trustworthy, so international standards are necessary, also for the proper calibration [8].

There will be an avalanche of data for which we will have to develop selection criteria with the help of knowledge management and biostatistics.

There are signs that the digital divide we talked about 10 years ago might become smaller, even in developing countries the use of cellphones for health purposes is increasing rapidly. Whether it will change the literacy divide rests to be seen, but it will most definitely cause a, temporary, generational divide. It will take the time till the current generation of children will be grandparents before the digital world will really be around us. Till that time we will have to put a lot of effort in making people aware. That is both patients and caregivers.

Due to time limits, I have not tackled the problems of privacy. It is estimated that a citizen of the UK can be found in 700 different databases. And now we are adding our medical and health information to it. Technology increases the possibility to link these databases. And the view on privacy from my parents generation, where neighborhood gossip was about the worst that could happen, is far different from privacy perceptions of the youngest generations that adapted to or grow up with facebook [9].

6 Conclusion

Whilst writing this presentation, I realized that even after 6 years, our core business as ICMCC is awareness:

Access to structured information and
Worldwide
Availability of data;
Realization through
Education, considering
National culture and tradition; thus obtaining
Emancipation,
Security and
Safety of the patient [10].

References

1. ICMCC Record Access Portal. [Online]. Available from: URL:http://recordaccess.icmcc.org/
2. Fisher B, Fitton R, Bos L. WHO recommendation on record access (draft). In: Bos L, Blobel B, editors. Medical and Care Compunetics 4. Amsterdam: IOS Press; 2007. p. 311–5. Available from: URL: http://recordaccess.icmcc.org/category/WHO/.

3. Bos L, Marsh A, Carroll D, Gupta S, Rees M. Patient 2.0 Empowerment. In: Arabnia HR, Marsh A, editors. Proceedings of the 2008 International Conference on Semantic Web & Web Services SWWS08. 2008. p. 164–7.
4. Capstrat. Health care information—Where do you go? Who do you trust? [Online]. Available from: URL:http://www.capstrat.com/elements/downloads/files/health-care-information-where-do-you-go-who-do-you-trust.pdf. Accessed 2010 Sep 3.
5. Hannan A, Webber F. Towards a Partnership of Trust. In: Bos L, Blobel B, editors. Medical and Care Compunetics 4. Amsterdam; IOSPress 2007. p. 108–16. Available from: URL: http://www.icmcc.org/pdf/recordaccess/hannan.pdf
6. HIStalk Interviews Don Holmquest, MD, JD, PhD, President and CEO, CalRHIO. 5 May 2008. [Online]. Available from: URL: http://histalk2.com/2008/05/05/histalk-interviews-don-holmquest-md-jd-phd-president-and-ceo-calrhio/ accessed 2008 May 6.
7. Doctors find errors in tenth of SCRs. Smart Healthcare. 20 July 2010. [Internet] Available form: URL:http://www.smarthealthcare.com/south-birmingham-scrs-tenth-errors-lmc-20jul10 Accessed 21 July 2010.
8. Padfield PL. The case for home monitoring in hypertension. BMC Med. 2010 Sep 27;8(1):55. Available from: URL:http://www.biomedcentral.com/content/pdf/1741-7015-8-55.pdf. Accessed 13 Oct 2010
9. Garner J, O'Sullivan H. Facebook and the professional behaviours of undergraduate medical students.Clin. Teach. 2010;7(2):112–5. Available from: URL:http://onlinelibrary.wiley.com/doi/10.1111/j.1743-498X.2010.00356.x/full. Accessed 2010 Sep 10.
10. Bos L. Medical & Care Compunetics and Patient Safety. Speech at BioMedea Conference. Stuttgart. Available from: ULR: http://www.icmcc.org/2005/09/25/biomedea-conference-2005/. Acessed 23 Sep 2010.

Knowledge Management and E-Health

Rajeev K. Bali, M. Chris Gibbons, Vikraman Baskaran and Raouf N. G. Naguib

Abstract Knowledge management (KM) has made a significant impact on the global healthcare sector. However, it is important to address the link between knowledge, information and engineering. This paper discusses how concepts from the established KM field can be applied to the area of e-Health. Explicit and tacit modes of knowledge transfer are presented and discussed. We conclude by advocating "tacit-to-tacit" knowledge transfer as the most useful method to overcoming knowledge gaps in various clinical and healthcare settings.

Keywords Knowledge · Tacit · Explicit · Knowledge management · Healthcare · Urban health · Breast screening · Lifetime health record

1 Introduction

Like many abstract entities, the concept of knowledge also differs according to the context in which it is interpreted—it may be as an object that might be identified, created, captured, stored and accessed or as a process that has a strong

R. K. Bali (✉) and R. N. G. Naguib
Biomedical Computing and Engineering Technologies (BIOCORE), Applied Research Group, Coventry University, Coventry, UK
e-mail: r.bali@coventry.ac.uk

M. C. Gibbons
Johns Hopkins Urban Health Institute, Baltimore MD, USA

V. Baskaran
Research Lab for Advanced System Modelling, Ted Rogers, School of Management, Ryerson University, Toronto ON, Canada

Commun Med Care Compunetics (2011) 1: 61–74
DOI: 10.1007/8754_2010_2

Published Online: 17 November 2010

human-centric orientation in culture, trust, beliefs and values [1, 2]. Healthcare institutions have now realized the true potential of knowledge and are trying, in various ways, to move forward in the new "knowledge era". Whether knowledge is interpreted as an object or as a process [3], the business world has also accepted that knowledge is the way forward [4], especially in organizations where the prime deliverable is service-oriented (such as healthcare). Knowledge richness makes healthcare the most receptive domain for KM-based improvements. Due to the fact that a large volume of healthcare knowledge is being lost because of its tacit-bias, even the smallest effort for managing this tacit knowledge can result in huge resource savings [5].

2 Knowledge Management

The term "knowledge" is often misunderstood and misinterpreted as it exhibits both physical and mental characteristics [1, 6]. All organizational domains, including healthcare, have embraced the concept of managing knowledge as the primary way of effective management for the future [4]. Domains predominantly focused in the aspect of service provision are more directly influenced and affected by "right or wrong" choices made towards implementing knowledge management (KM). Service-based domains (such as healthcare) are more influenced by these choices as such domains are "knowledge rich". They are therefore prone to become breeding grounds for the creation of knowledge gaps (often referred to as "islands" or "silos"). Eventually, this can lead to knowledge deprivation at the required point of delivery, ultimately losing the competitive edge [7, 8].

The fact that modern day organizations are facing a deluge of data and information whilst simultaneously lacking knowledge is very well documented [9, 10]. Technological innovations relating to workflow and such technologies as groupware systems have brought about a radical transformation in the way organizations can interact both internally and externally. These new ways of collaboration have resulted in organizations being inundated with information to an unprecedented degree resulting in data and information overload [10–12]. It has been said that knowledge is an inherent characteristic of the human brain [13]. It has opened a new vista for replicating this ability to a degree through techniques as artificial intelligence (AI). Despite such technological breakthroughs, the ability of machines to imitate human cognition is very limited. One of the promising areas for AI is in prediction [13] but even this is when there are a number of factors that are involved and a suitable training data set is available [14].

Technologies such as data integration, document and content management (support applications enabling users to have personalized access to the organizational knowledge base) continue to grow at an exponential rate [15]. This has implications for decision makers across all sectors including healthcare, as they have to deal with large amounts of data [16]. One of the major challenges that face healthcare managers is how to make effective decisions based on the data at hand. It is acknowledged that the selection of a particular direction is both constrained and influenced by the availability of data, the ability to transform data into information and then to make recognition of it by deriving knowledge from information. Practitioners then have to decide on how best to effectively transfer this knowledge on an organization-wide basis. KM tools and techniques are defined by their social and community role in the organization by way of:

- The facilitation of knowledge sharing and socialization of knowledge (production of organizational knowledge)
- The conversion of information into knowledge through easy access, opportunities of internalization and learning (supported by the right work environment and culture)
- The conversion of tacit knowledge into 'explicit knowledge' or information, for purposes of efficient and systematic storage, retrieval, wider sharing and application.

For the healthcare setting, knowledge deprivation can result in a loss of resources, making the care process inefficient [5]. Confusion still exists in differentiating knowledge from information [4]. In most cases, data is the point of origin for new knowledge creation within a typical knowledge cycle. Data can be viewed as a representation of unprocessed numbers, text and so forth. When given *context*, this same data is transformed into information. This same information, when assimilated by the cognitive environment recognizes similarities, connections or patterns in the information, paves the way to knowledge creation.

2.1 Types of Knowledge

Knowledge can be broadly classified as *tacit* when it is within the cognitive environment (e.g. a human brain) or *explicit* [17, 18], when the tacit knowledge is expressed for sharing (accompanied by the context of the created knowledge) through various tools [19]. Until now, managing knowledge has been specifically focused on the explicit type since that is often the only method available for sharing knowledge [20]. The irony of KM is that more importance is often given to technology-based aspects, rather than the more human-centric aspects [20]. The healthcare domain not only provides challenging opportunities for managing knowledge but also is one of the domains where it is often most poorly understood and deployed.

This predicament is slowly being addressed as more and more KM focused projects are initiated and professionals with better understanding of KM are being involved [21]. Any healthcare domain, whether primary care or secondary care in nature, relies on a lot of data and information flows. Similarly, modern day healthcare environments should also provide knowledge sharing through conducive and well established channels [22]. Knowledge gaps can be circumvented through removing bureaucracy and formal channels. This would pave the way to leveraging knowledge which has been created within the healthcare domain, thus increasing the overall efficiency in making well informed and correct clinical decisions [23]. The will to share knowledge has to originate from the human mind no matter what technological tools are available; the inclination to do so still has to be spontaneous and forthcoming. This aspect may be mapped to socio, cultural and moral ethics of the human beings involved in knowledge sharing [19]. In healthcare, there is no one "silver bullet" solution that can successfully address the KM issues; rather, the whole organizational environment has to be aligned to encourage knowledge creation and knowledge sharing [5]. Nonaka's knowledge creation company concept and the SECI model [2, 24] cannot be expected to justify the complex nature of modern day KM.

A radical new approach, focused on maximum knowledge sharing at a tacit-to-tacit level [19] would be an ideal beginning for the current KM challenges in healthcare. All projects in healthcare should be part of a KM initiative and all challenges should be approached through a knowledge-based solution and perspective. This would provide a holistic approach and a successful strategy to enable innovation and success at all levels [19]. We now present three case study examples which illustrate how knowledge-based initiatives can be used to foster effective e-Health solutions.

2.1.1 Example #1: Urban Health Knowledge

Urban health is a multidisciplinary and multisectoral approach to promoting public health and individual health in the urban setting. The field of urban health is concerned with the determinants of health and diseases in urban areas and with the urban context itself as the exposure of interest [25]. The complexity of urban health makes it suitable for focused research and examination. The multidisciplinary nature of urban health encompasses public, private and non-governmental sectors including public health, urban planning, social work, education, engineering, architecture, law, media, food and agriculture, community development, environmental protection, transportation, economics amongst several others [26].

Recent advances in the computer sciences and information technology fields have spawned several methodological advances in the biological and molecular sciences (e.g, DNA chip technology and microarray analysis), enabled quantum leaps in molecular and submolecular medicine, and catalyzed the emergence of whole new fields of study such as proteomics, phenomics, nutrigenomics, and pharmacogenetics. Perhaps, in like manner, with the emergence of eHealth, the behavioral and population sciences may be on the verge of a similar information technology-based scientific revolution. New eHealth solutions may soon permit the real-time integrative utilization of vast amounts of behavioral-, biological-, and community-level information in ways not previously possible.

Behavioral algorithms and decision support tools for scientists could facilitate the analysis and interpretation of population level data to enable the development of "community (population) arrays" or community-wide risk profiles, which in turn could form the foundation of a new "Populomics." [27]. The term Populomics has emerged from the synthesis of the Population sciences, Medicine and Informatics [27, 28]. Populomics is defined as an emerging discipline focused on population level, transdisciplinary, integrative disease/risk characterization, interdiction and mitigation that rely heavily on innovations in computer and information technologies. Populomics seeks to characterize the interplay of socio behavioral pathways and biophysiologic and molecular mechanisms which work across levels of existence, to impact health particularly, at the population level.

KM embodies organizational processes that seek synergistic combination of data and information processing capacity of information technologies, and the creative and innovative capacity of human beings. Populomics, a confluence of three seemingly disparate concepts (population science, medicine and informatics) has parallels with the growing field of KM, itself a blend of people, process and technology [28]. KM principles and methodologies provide researchers and practitioners (concerned with health impacts associated with the urban environment) valuable analytic tools and a systematic approach to the integration of different types of data to enable novel knowledge-based insights and spur scientific advances in urban health.

It has been argued that such improvements and improvements in disciplines as supposedly diverse as organizational behavior, ICT, teamwork, artificial intelligence, leadership, training, motivation and strategy have been equally applicable and relevant in the clinical and healthcare sectors as they have been in others. Clinicians and managers have used many of these disciplines (in combination) many times before; they may have, inadvertently and partially, carried out knowledge management *avant la lettre* [29]. Understanding and disentangling the myriad determinants of disease, particularly within the context of urban health or health disparities (inequalities in health), requires a transdisciplinary approach. Transdisciplinary approaches draw on concepts from multiple scientific disciplines to develop integrated perspectives from which to conduct scientific investigation and provide needed care. Attempts to organize and understand complex bio-socio-behavioral systems have led some researchers to *Chaos Theory* and *Complexity Theory* as constructs to facilitate the understanding about health and its relationship to diverse processes and outcomes [30–33]. In reality though, it is likely that these approaches are beyond the practical usefulness of many clinicians and scientists. Recently, elaboration of the Sociobiologic Integrative Model (SBIM) has been advanced as a theoretic construct to facilitate the integration of knowledge from many different fields [34]. Utilizing the SBIM along with the principles of KM may offer health disparities researchers and clinicians providing care in the urban environment significant promise towards the quest to improve Urban Health and eliminate health disparities/inequalities.

2.1.2 Example #2: Breast Screening Service

Breast cancer is the most common cancer in women with over 40,000 women being diagnosed with the disease each year in the UK [35]. Any information related to the breast can largely affect a women's consciousness and a threat of breast cancer will have varying impacts on women psychology. Typically breast cancerous cells originate in the mammary glands (lobules) or in the ducts connected to these glands or in other tissues around these glands [36]. When in close proximity to the lymphatic system, these cells can result in being carried to other organs of the body. This subsequently results in cancerous growth in that organ and is described as metastatic breast cancer [36]. Although many causes had been identified for breast cancer, the knowledge of finding a cure is still not within the reach of modern medicine.

Breast cancer should ideally be diagnosed at the earlier stages of its development. Possible treatments include removing or destroying the cancer cells to avoid the spread of the affected cells. Breast self examination (BSE) is an effective and non-intrusive type of self diagnosis exercise for checking any abnormalities/lumps in the breast tissue. Unfortunately this greatly depends on the size of the lump, technique and experience in carrying out a self examination by the woman. An ultrasound test using sound waves can be used to detect lumps but this is usually suited for women aged below 35 owing to the higher density of breast tissue [36]. Having a tissue biopsy via a fine needle aspiration or an excision is often used to test the cells for cancer. These tests are mostly employed in treatments or post-treatment examination and as second rung diagnostic confirmation methods. Performing a computed tomography (CT) or a magnetic resonance imaging (MRI) scan would result in a thorough examination of the breast tissue but this technique is not favored due to reasons including that it may not be economical, needs preparation, noisy, time consuming and images may not be clear.

Breast Screening Programme

Mammography is a technique for detecting breast tissue lumps using a low dosage of X-ray. This technique can even detect a 3 mm sized lump. The X-ray image of the breast tissue is captured and the image is thoroughly read by experienced radiologists and specialist mammogram readers. Preliminary research suggests that women aged 55 and above are more susceptible to breast cancer; mammography is more suited to the women aged 55 and above (due to the lower density of breast tissue) [37]. Even though mammography has its critics—mainly due to its high rate of false positives and false negatives [38]—it has still become the standard procedure for screening women by the National Health Service (NHS) National Breast Screening Programme UK [39]. Mammography is the best and most viable tool for mass screening to detect cancer in the breast at an early stage [40]; however, the effectiveness of diagnosis through screening is directly dependent on the percentage of women attending the screening programme. The NHS Breast

Screening Programme, catering to the entire eligible women population is funded by the Department of Health in the UK and is the first of its kind in the world. It covers nearly four million women and detected more than 13,000 cancers in the screened population for the year 2005 [41]. Currently the screening programme routinely screens women between the ages of 50 and 70, and employs two views of the breast, medio-lateral and cranio-caudal. The lack of breakthroughs in finding a definitive cure means that preventive medicine is the only viable alternative to reduce deaths due to breast cancer.

The UK NHS National Breast Screening Programme (NBSP) is unique as it provides free breast screening for the female population aged between 50 and 70 at a national level [39, 42]. The recent increase of the upper age limit from 63 to 70 for screening and making a two-view mammogram mandatory has greatly increased the efficiency of benign or malignant tumor detection. The NBSP currently runs a massive screening programme catering to almost two million eligible women across the UK [42]. This programme runs on a call/recall cycle which screens all eligible women in a 3 year interval. The information published by the UK Government Statistical Service (NHS Health and Social Care Information Centre) in its Community Health Statistics report for the year 2006 agrees that, for the past ten years since 1995, the uptake has remained constant at around 75%. The number of non-attendees has been significantly increasing and has reached half a million. Simple projection of this data submits that nearly 4,000 cancer incidences would not have been diagnosed. Even if a small percentage of these non-attendees could be made to attend, it would result in the saving of significant lives. Indirectly we can also infer that, despite focused efforts on these non-attendees for the past ten years, there was no real effect on their attendance. Moreover, early stage cancer detection would have a huge impact in reducing cancer related deaths. From these facts and data, we see that the primary concern is to reduce non-attendance [43].

These challenges can be addressed by a resource saving strategy which has better healthcare at its core. The first component in the proposed strategy is related to knowledge via AI, employing neural network (NN) algorithms. This strategy also includes a service oriented architecture (SOA) to deliver the envisaged knowledge as the second component. This research proposes to unify the existing national breast screening computer system (NBSS) software onto a single platform and create prototype software component based on open source technologies. The proposed prototype software would be automated to produce the pre-processed data and eventually normalize the data for AI (neural network) assimilation. These activities would be performed sequentially without human involvement for repeatability, reliability and accuracy. The Java based attendance prediction by artificial intelligence for breast screening (JAABS) model itself would be simulated on the open source technology platform. This model incorporates all additional transformations occurring within the screening process (including the change in the screening upper age limit). The prototype framework proposed will incorporate the AI model for creating a list of predicted non-attending women. The prototype combines the demographic data pertaining to the non-attending women and information related to her general physician (GP) as a messaging package. This package triggers the generation of an electronic message based on the Health Level 7 (HL7) version 3 standards and utilizes SOA as the message delivering technology.

Implications for KM practice

The objective of this work was to identify the challenges which are being faced by the UK NHS' national breast screening programme and find approaches to alleviate these impediments and eventually reduce mortality due to breast cancer. Based on JAABS algorithm's negative prediction value (for the first to sixth episode) the number of non-attendees correctly predicted a priori to screening date would be at least 42 women for every 100 screening non-attendees. When such knowledge is shared with the GPs (with whom the women are registered) can initiate interventions. Such interventions can educate the non-attending women and clarify their attitudes and beliefs. The expected outcome is that the women commits to a positive informed decision, which would culminate in attending the screening appointment. This work not only confirms that breast screening attendance can be predicted through an automated software solution, but also can be leveraged to increase screening attendance by employing emerging KM tools and techniques. This research work draws its strength from such KM tools and techniques. This work is also one such initiative addressing the NHS' breast screening attendance through efficient KM methodologies. A 25% success in GP interventions will result in saving more than 350 women's lives per year. Even if one women's life can be saved by our approach, this approach can be deemed as a success. The new bespoke software prototype, incorporating JAABS algorithm can be easily converted and integrated into the NBSS.

2.1.3 Example #3: Malaysian Lifetime Health Record

In trying to propel Malaysia into a "developed nation" status, the Government of Malaysia has formulated vision and mission statements such as the Malaysian Vision 2020 as well as the Malaysian Healthcare Vision. In trying to actualize both these visions into reality, the government and the Ministry of Health of Malaysia (MOHM) have started preparing itself through the *Integrated Telehealth Initiative* where the lifetime health record (LHR) is used as the basis for continuous care. The LHR correlates each episode of care for an individual into a continuous health record. It is the summarized health record of every individual compiled from their electronic medical records. These records refer to a patient's electronic medical records that are cumulatively derived from the clinical support system (such as clinical information system, laboratory information system, pharmacy information system and patient management system) and it can be collected and gathered from the various spectrums of health information systems and healthcare levels.

The Malaysian public healthcare system is structured in a hierarchical pyramid-based concept. At the base of the pyramid is a broad array of primary healthcare services (such as health centers, polyclinics, mobile clinics and maternal and child clinics) spread throughout the country. The next level consists of district hospitals in every one of 120 districts, feeding into state general hospitals in each state capital. At the top of the pyramid lies the Hospital Kuala Lumpur which is the national tertiary reference centre which provides specialist and super specialist services for the nation [44, 45]. In term of the physical healthcare facility setup, the healthcare premises are normally developed within the area where the most population is living [46]. Since 1998, 95% of the population were living within a 5-kilometre radius of the nearest healthcare facility. This setup enables patients from anywhere in the country to be referred to the appropriate hospital, to access and visit several healthcare facilities through a nationwide network of clinics, hospitals and other health programs in a convenient manner.

Background of Outpatient Clinics

The outpatient clinics department administratively reports to the Family Health Division of Public Health Services of MOHM. Healthcare services are provided through various health centers and community polyclinics strategically located in the most populated areas in the district. The health centers and community polyclinics comprise the first level of service made available to the community. The services provided are comprehensive at this level, essentially comprising maternal health, child health, acute care of diseases, chronic care of diseases, mental health, geriatric care, community-based rehabilitation, well person services and health promotion. These services are provided as outpatient treatments. In order to support such services, there are laboratory services and also radiological services. The pharmaceutical services are also provided in-house. With these comprehensive services and workflows, it was a significant reason that the out-patient clinic department was selected as a case study organization for advancing knowledge on the case under study. The selected organization will cover most of the processes of consultation and medical diagnosis workflow in outpatient clinic. The healthcare setting provides relevant evidences and extensive information about patient demographic and clinical data that would contribute in developing the LHR components and structure.

Based on inputs and evidences obtained from the case study, the LHR was divided into three components: (1) patient master information, (2) health condition summary and (3) episode summary.

Patient Master Information (PMI)

Patient master information comprises of administrative records and the required information to identify and distinguish the patient across healthcare facilities and levels. It often used to locate patient identifier, including patient demographic and the related health administration information. The patient master information comprises the following set of information including demographic record, next of kin record, birth record, family health record, medical insurance record, employment record and organ donor record. It was noted from the findings of the primary data collection, the analysis of patient demographic information to be viewed by doctors during consultation) that the patient demographic information is highly required during consultation and medical diagnosis workflow. The demographic information is indicated as compulsory due to the fact that the demographic information is a key identifier for the patient. The administrative records (examples are next of kin record, birth record, family health record, medical insurance record, employment record and organ donor record) could be included in the patient information as optional records.

Health Condition Summary

The health condition summary comprises of records, which summarize the illness and wellness condition of the patient. Each condition has a status indicator to indicate whether the condition is active or inactive. This summary of a patient's condition will enhance the continuity of care by providing a method for communicating the most relevant information about a patient and providing support for the generation of LHRs [47, 48]. It was noted from the primary data collection that the first step of patient care or treatment was that the doctor will gather information about the patient's current health status. Here, many types of information were collected about the patient and placed in the patient's health record. By giving the latest health condition summary of the patient at the beginning of a first doctor-patient encounter, the accuracy, quality, safety and continuity of care would be given. The health condition summary component could be added and enhanced in future and the set of information given below are the initial information revealed from the primary data collection and system analysis. The health condition summary comprises such information as chronic disease record, allergy record, immunization/vaccination record, social history record, surgical medical procedure record, disability record and obstetric record.

Episode Summary

The episode summary is comprised of data for a particular episode or visit. If required, it provides the necessary data for reference to the source of the information where details of the episodes are stored. It comprises the following information; episode record, encounter record, symptoms record, diagnosis record, lab test record, radiology record, medication record, vital sign record and health plan record. The LHR components defined above provide the conceptual structure of LHR information and it is envisaged that the many LHRs could be collected and generated continuously from telehealth applications and various health information systems. The integrated telehealth system, via its role as the premier database repository for patient's LHR in the country, will function as the health knowledge platform from which a variety of research and development activities within the healthcare domain can be launched. These research and development initiatives will spearhead the development and establishment of initiatives which will produce innovative healthcare products and services of national and international significance.

Implications for KM practice

The KM implications and practice in this case are varied. The integrated telehealth system, by way of its role as the premier database repository for patient's LHR in the country, will function as the health knowledge platform from which a variety of research and development activities within the healthcare domain can be launched [49]. These initiatives will spearhead the development and establishment of initiatives which will produce innovative healthcare products and services of national and international significance. In terms of health group data services, the LHR's standardized data sets would ensure effective mining (data, information and knowledge). The LHR repository would be a premier health database repository in the country and would become an important source for health knowledge data mining purposes [50]. From judicious use of standardized KM applications, personalized health plans can be formulated [51].

3 Conclusion

This chapter has presented several knowledge-based cases which exemplify the efficacy of information and communication technologies for effective eHealth solutions. Current knowledge-based initiatives concentrate heavily on knowledge transfer in tacit-to-explicit modes which do not yield the expected outcomes. We argue that a more tacit-focused mode of knowledge transfer would achieve a more effective transfer of knowledge. Judicious and integrative use of novel concepts (such as Populomics for urban health) will help enable these knowledge-based constructs. Any future knowledge-based implementations in the three examples provided would do well to orient efforts on sharing knowledge in tacit-to-tacit manner in order to avoid knowledge gaps and leverage all of the available, multidisciplinary, resources.

Acknowledgments The Breast Screening and Malaysian Lifetime Health Record case studies appear courtesy of the Biomedical Computing and Engineering Technologies (BIOCORE) Applied Research Group, Coventry University, UK (www.coventry.ac.uk/biocore/). The authors appreciate the opportunity to contribute to this volume in memory of the Late Prof. Swamy

Laxminarayan. Two of the authors (Bali and Naguib) would like to acknowledge the seminal work of Prof Laxminarayan in the fields of biomedical engineering and clinical management—his constant encouragement, guidance and friendship to us will be sorely missed.

References

1. Miller FJ. I = 0 (Information has no intrinsic meaning). Inf Res Int Electron J. 2002;8:1. http://informationr.net/ir/8-1/paper140.html. Accessed 15 May 2009.
2. Nonaka I, Takeuchi H. The knowledge-creating company. Oxford: Oxford University Press; 2002. p 61–94. ISBN: 0-19-509269-4.
3. McElroy MW. The new knowledge management. Burlington: Butterworth-Heinemann Publications. KMCI press; 2003. p 13–18. ISBN: 0-7506-7608-6.
4. Paul MH, Kimble C. The duality of knowledge. Inf Res Int Electro J. 2002;8:1. http://informationr.net/ir/8-1/paper142.htm. Accessed 21 May 2009.
5. Burns S. Knowledge management in healthcare: succeeding in spite of technology. Business J. 2003;32–36. http://www.hmc-benchmarks.com/about/SBurns_ASFMI.pdf. Accessed 25 Apr 2006.
6. Breen M. Information does not equal knowledge: theorizing the political economy of virtuality. J Comput Mediat Commun. 1997;3:3. http://www.ascusc.org/jcmc/vol3/issue3/breen.html. Accessed 25 Aug 2009.
7. Rodrigo BC. Using information technology to support knowledge conversion processes. Inf Res Int Electron J. 2001;7:1. http://informationr.net/ir/7-1/paper118.html. Accessed 05 June 2006.
8. Smith L, Preston H. Information management and technology strategy in healthcare: local timescales and national requirements. Inf Res Int Electron J. 2000;5:3. http://informationr.net/ir/5-3/paper74.html. Accessed 5 Jan 2007.
9. Liebowitz J, Beckman T. Knowledge organizations: what every manager should know. Delray Beach: St. Lucie Press; 1998.
10. Sieloff C. If only HP knew what HP knows: the roots of knowledge management at Hewlett-Packard. J Knowl Manag 1999;3(1):47–53.
11. Cothrel J, Williams R. On-line communities: helping them form and grow. J Knowl Manag 1999;3(1):54–60.
12. Drucker P. The coming of the new organization. Harvard business review on knowledge management, Boston: Harvard Business School Press; 1998. p 1–19.
13. Baskaran V, Bali RK, Arochena H, Naguib RNG, Wheaton M, Wallis M. Improving uptake of a breast screening programme: a knowledge management approach for opportunistic intervention, In: Bos L, et al., editors. Medical and care compunetics 3, Amsterdam: IOS Press; 2006. p 191–7.
14. Protti DJ. The application of computer technology in GP surgeries is beginning to have positive effects on chronic disease management. Prof. DJ Protti's World View Reports, NHS CFH Press office. 2005. http://www.connectingforhealth.nhs.uk/newsroom/worldview/protti5. Accessed 20 Apr 2005.
15. Bali RK, Feng DD, Burstein F, Dwivedi A. Advances in clinical and health-care knowledge management. IEEE Trans Inform Tech Biomed. 2005;9(2):157–61.
16. Bali RK, editor. Clinical knowledge management: opportunities and challenges. Hershey: IGP; 2005.
17. Koskinen KU. Evaluation of tacit knowledge utilization in work units. J Knowl Manag. 2003;7(5):67–81.
18. Suliman H. Knowledge management: re-thinking information management and facing the challenge of managing tacit knowledge. Inf Res Int Electron J. 2002;8(1). http://informationr.net/ir/8-1/paper143.html. Accessed 5 Jan 2007.

19. Baskaran V, Bali RK, Arochena H, Naguib RNG, Dwivedi AN, Nahy NS. Towards total knowledge management for healthcare: clinical and organizational considerations. Proc of the IEEE Ann Int Conf of the Eng in Med and Biol Soc (EMBS) 2004; p. 3163–66.
20. Timo K. Knowledge management process model. Technical Research Centre of Finland, VTT Publications 2001;455. 101p. + app. 3p. ISBN: 951–38–5965–7/ISSN: 1235–0621.
21. Open Clinical. The medical knowledge crisis and its solution through knowledge management—White Paper—(DRAFT- v.3-15). 2000. http://www.openclinical.org/docs/whitepaper.pdf. Accessed 5 Jan 2007.
22. Augier M, Shariq SZ, Vendelø MT. Understanding context: its emergence, transformation and role in tacit knowledge sharing. J Knowl Manag. 2001;5(2):125–36.
23. Ellingsen G. The role of trust in knowledge management: a case study of physicians at work at the university hospital of northern Norway. Inform Sci J. 2003;6:93–207. http://inform.nu/Articles/Vol6/v6p193-207.pdf. Accessed 10 June 2009.
24. Kikawada K, Holtshouse D. Managing industrial knowledge. Nonaka I, Teece D, editors. London: Sage; 2001. p 306–314 ISBN: 0-7619-5498.
25. Galea S, Vlahov D. Urban health: evidence, challenges and directions. Ann Rev Public Health. 2005;26:341–65.
26. Abrams DB. Applying transdisciplinary research strategies to understanding and eliminating health disparities. Health Educ Behav.2006;33(4):515–31.
27. Gibbons MC. Populomics. In: Bos L, Blobel B, Marsh A, Carroll D, editors. Medical and care compunetics 5. Amsterdam: IOSPress; 2008. p. 265–8.
28. Wickramasinghe N, Bali RK, Lehaney B, Schaffer J, Gibbons MC. Healthcare knowledge management primer. New York: Routledge; 2009.
29. Bali RK, Dwivedi AN, Naguib RNG. Issues in clinical knowledge management: revisiting healthcare management. In: Bali R, editor. clinical knowledge management: opportunities and challenges. Hershey: IGP; 2005.
30. Garfinkel A, Spano ML, Ditto WL, Weiss JN. Controlling cardiac chaos. Science. 1992;257 (5074):1230–5.
31. Weiss JN, Garfinkel A, Spano ML, Ditto WL. Chaos and chaos control in biology. J Clin Invest. 1994;93(4):1355–60.
32. Olsen LF, Schaffer WM. Chaos versus noisy periodicity: alternative hypotheses for childhood epidemics. Science. 1990;249(4968):499–504.
33. Tidd CW, Olsen LF, Schaffer WM. The case for chaos in childhood epidemics. II. Predicting historical epidemics from mathematical models. Proc Biol Sci. 1993;254(1341):257–73.
34. Gibbons MC, Brock M, Alberg AJ, Glass T, LaVeist TA, Baylin SB, et al. The Socio-biologic integrative model: enhancing the integration of socio-behavioral, environmental and bio-molecular knowledge in urban health and disparities research. J Urban Health. 2007; 84(2):198–211.
35. Cancer Research UK. CancerStats incidence-UK. 2005. http://www.cancerresearchuk.org/aboutcancer/statistics/statsmisc/pdfs/cancerstats_incidence_apr05.pdf. Accessed 10 Aug 2005.
36. American Cancer Society Inc. Cancer reference information. 2005. http://www.cancer.org/docroot/CRI/content/CRI_2_4_1X_What_is_breast_cancer_5.asp. Accessed 10 Aug 2005.
37. Blanks RG, Moss SM, McGahan CE, Quinn MJ, et al. Effect of NHS breast screening programme on mortality from breast cancer in England and Wales, 1990–8: comparison of observed with predicted mortality. BMJ. 2000;321(7262):665–9.
38. Burton G. Alternative medicine. Washington: Future Medicine Publishing; 1997.
39. Forrest P. Breast cancer screening—a report to the health ministers of England, Scotland, Wales and Northern Ireland. London: HMSO; 1986
40. Medicine net. Breast cancer. 2002. http://www.medicinenet.com/breast_cancer/page3.htm. Accessed 10 Aug 2005.
41. NHS Review (2005) The breast screening programme annual review 2005. http://www.cancerscreening.nhs.uk/breastscreen//publications/2005review.html. Accessed 14 July 2007.

42. Cancer Research UK. Breast cancer factsheet. 2004. http://publications.cancerresearchuk.org/epages/crukstore.sf/en_GB/?ObjectPath=/Shops/crukstore/Categories/BrowseBySubject/BreastCancer. Accessed 18 Sep 2004.
43. Bankhead C, Austoker J, Sharp D, Peters T, et al. A practice based randomized controlled trial of two simple interventions aimed to increase uptake of breast cancer screening. J Med Screen. 2001;8(2):91–8.
44. Abu Bakar MA. Malaysia healthcare delivery system. Personal interview with MAK Ghani, Coventry University—PhD candidate. 2007.
45. Ariff KM, Teng CL. Rural health care in Malaysia. Aust J Rural Health. 2002;10(2):99–103.
46. Suleiman AB. The untapped potential of telehealth. Int J Med Inform. 2001;61(2–3):103–12.
47. Medical Record Institute. Continuity of care record (CCR)—the concept paper of the CCR—version 3. 2006. http://medrecinst.com/pages/about.asp?id=54. Accessed 2 Apr 2008.
48. Ministry of Health Malaysia. Telemedicine blueprint: Telemedicine flagship application. Government of Malaysia. 1997.
49. Harun MH. Integrated telehealth: what does it all mean? DTP Enterprise Sdn Bhd. 2002.
50. Ministry of Health Malaysia. Concept request for proposal for lifetime health plan: telemedicine flagship application. Government of Malaysia. 1997.
51. Mohan J, Raja Yaacob RR. The Malaysian telehealth flagship application: a national approach to health data protection and utilisation and consumer rights. Int J Med Inform. 2004;73(3):217–27.

Beyond the Fringe? Investigating the Boundaries of Healthcare

Bryan R. M. Manning

Abstract This chapter questions the constraints placed by the current boundaries of Healthcare delivery on the wider issues affecting the healthiness of the population at large, together with those whose problems do not fall within the scope of the traditional medical model. It examines how increasing complexity has fostered specialisation and with it the segmentation of care, together with its predominant focus on illness as compared with the attention given to wellness. Of particular concern is the somewhat ambiguous position of life-long disabling conditions that sit rather uncomfortably between these two opposing camps. Whilst this dichotomy opens up the issue of how damaged individuals can be helped to live as normal a life as possible, it also emphasises that health and wellness are more than not being ill! It enables a strong case to be made for extending the boundaries of care out to include coping with issues of sociological dysfunctionality as part of long-term well-being.

Keywords Healthcare delivery · Boundaries · Wellness · Population health

1 Introduction

The concepts of health and healing have become so intertwined with the passage of time that they have almost become synonymous. This is somewhat unfortunate as the overall concept of health centres on well-being, as opposed to the healing processes. Indeed this inversion is carried forward into the title of the UK National

B. R. M. Manning (✉)
School of Engineering and Computer Science, University of Westminster, London, UK
e-mail: bryan.manning@btinternet.com

Commun Med Care Compunetics (2011) 1: 75–84
DOI: 10.1007/8754_2010_6

Published Online: 22 March 2011

Health Service, which in reality is predominantly a remedial 'healing service'—not as some have unkindly suggested an 'illness service'!

This overarching concept of health was defined on its creation by WHO defined as:

> a state of complete physical, mental, and social well-being and not merely the absence of disease or infirmity [1].

However almost inevitably countering national disease burdens were been given priority as the key step toward achieving the wider target of increased well-being.

However as a consequence of finite resources, there has been a trend to split off the sociological component and focus separately on the medical aspects. Even then this tendency to compartmentalize in the face of complexity has driven a further wedge between the physiological and psychiatric domains.

So far the response of the clinical care professions to ever increasing complexity has been to specialize—generating an expanding range of disciplines and agencies that tend to operate in domain silos. This in turn adds further layers of complexity through domain separation of operational models—most importantly accompanied by a lack of any overarching shared strategic model.

The result is that outcomes are domain specific and do not deal with the individual as a composite whole—and can all too easily lead to domain experts inadvertently undoing each other's efforts or merely shunting the core problems about to no effect or worse. This is also not helped by the almost universal lack of any shared interdisciplinary plans or clinical pathways (Fig. 1), which would at least go some way to highlighting treatment interdependencies.

Whilst indicating some of the areas where basic improvements could be made, the key problem remains. Namely, how to radically re-model services to provide a more unified care approach spanning the physical, mental, and social domains from the combined context of well-being and healing perspectives.

Some initial moves are being made in this direction in response to the elderly demographic 'time bomb' by focusing on helping the elderly to maintain their independence at home for as long as possible. However even though this is rather late in the day it is at least a start, albeit limited by current constraints on finance as well as those of human and technological resources.

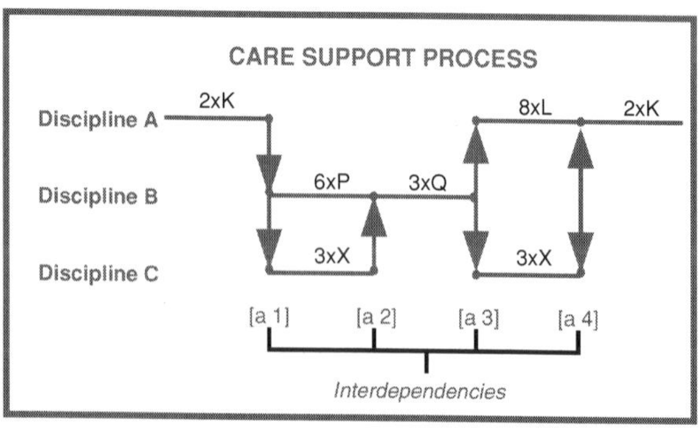

Fig. 1 Multi-disciplinary care pathway

Unfortunately in the latter stages of life we are imperceptibly beset by increasing frailties of mind, body and circumstance, each of which may need some level of help and support. However what is needed is coordinated input across the spectrum of individual need, appropriately channeled from different disciplinary sources, rather than the current well-meaning but rather haphazard multi-service/agency delivery.

From this it would seem reasonable to infer that support for through-life preventive action to help maintain well-being and independence through the life course should radically reduce exposure and hence subsequent demand on remedial services—albeit at a heightened medium term cost.

2 Impairment

Ageing is frequently seen by the younger generations as an increasingly fraught medical condition, in reality it is one of slow degeneration of functionality based on impairments and life-changing events that stretch back to earlier times in an individual's life [2].

However both the triggers and ramifications of these events are not solely confined to the latter stages of life, but span the complete life course from birth to death, and can have impacts across the whole range of physical, mental and sociological well-being.

These impairments can range from the mildly inconvenient to the catastrophically disabling (Fig. 2), with the common feature that remedial action is centred on helping the sufferer to cope or adapt to conditions that are unlikely to yield curative measures.

For low level or early stages of impairment sufferers generally chose to deal with their problems either alone or with some measure of advice and guidance. However as time passes 'creeping' impairment tends to set in as the level of

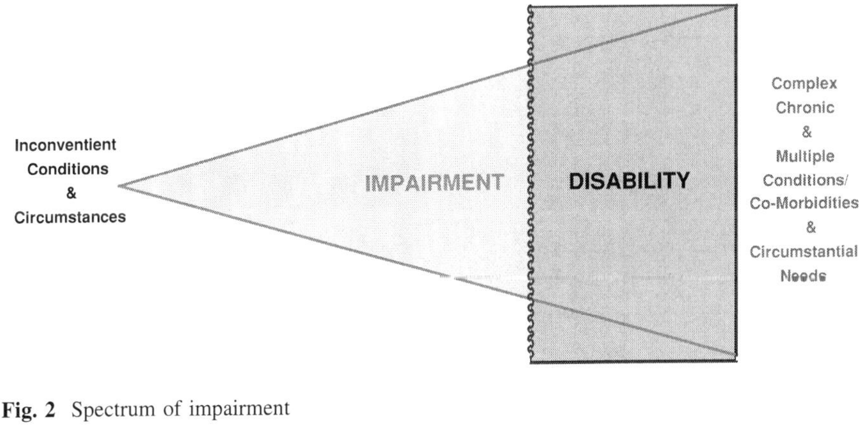

Fig. 2 Spectrum of impairment

inconvenience increases to the point that it becomes significantly disabling—especially as this can trigger other effects and resultant needs.

The problem is that this is likely to involve aspects of all three areas of well-being, which without comprehensive coordinated effort will accentuate matters and lock the sufferers into a downward spiral of dependency.

However the trouble is that the resultant needs profiles do not fit neatly into primary service segments, multiple agencies or myriad specialties that potentially can be involved. Moreover the lack of any overarching strategic responsibility or service coordination inevitably results in patchy and often poor quality support and care delivered to sufferers, coupled with considerable waste of effort and resources—however well meant.

Whilst action to put in place improvements to create more comprehensive and effective support mechanisms would obviously be welcome, so too would preventive measures to help people to avoid or delay the impacts of low level impairment for as long as possible (Fig. 3).

Broadly this divides between avoidance through lifestyle behavioral change, or coping strategies. The preventive route centers on a range of lifestyle improvement elements that include exercise, diet and personal skills development. By contrast the self-motivated option relies on a combination of innate cognitive abilities and problem solving skills.

Ultimately both alternatives can benefit from support from various "smart" tele-health technologies that can provide appropriate guidance and advice, and whose acceptance and ultimate success is dependent on ease of access and assimilation, together with obvious relevance to those concerned.

Since these services centre on achieving behavioral change, they incorporate a variety of both psychological and psychotherapeutic input, whilst focusing on a wide range of psycho-social and socio-economic issues. Although there is a considerable

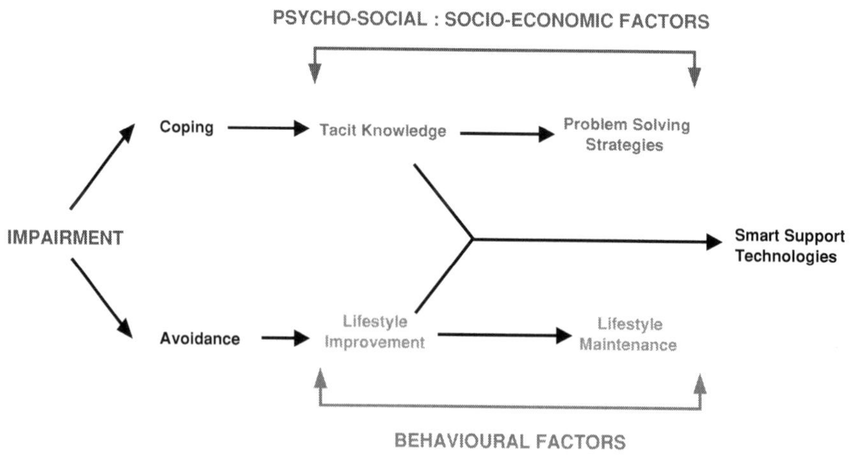

Fig. 3 Optional responses to impairment

body of knowledge that can be readily deployed to help improvement in well-being [3], by contrast relatively little knowledge appears to have been collated for coping strategies in these domains in comparison with that for physiological conditions.

3 Classification of Needs

From a generic point of view change is driven by determination of specific needs or requirements that lead via a planning stage through to implementation and subsequent maintenance. The success of any such mission must hinge on correctly identifying the root causes and factors exacerbating the scale of the problem that is presented, and the criteria against which the outcome can be judged.

In clinical terms this process chain crudely comes down to completing a diagnosis, assessing the treatment options, and implementing the most appropriate intervention pathway as treatment processes to its ultimate outcome. Whilst the medical profession has a classification system for codifying diseases and disorders—ICD 10 [4], it has yet to succeed in achieving a similar approach for the ever-evolving host of clinical pathways.

By contrast the psycho-social and socio-economic domains are less well served, with the exception of the WHO classification of Functioning, Disability and Health—ICF [5]. Although its prime focus is on functioning and disability, it sets this in the context of societal and environmental factors, and is at pains to point out that 'a decrement in health … [implies] … some degree of disability'—viz. impairment.

Whilst ICF provides valuable insights and inputs from the perspective of personal functioning, it does not pick up the many life events [6] that trigger major psycho-social or socio-economic dysfunctions. The lack of a comprehensive classification system that spans the mental health and community care fields and codifies its dysfunctions puts them at a significant disadvantage in attempting to resolve the service complexities involved.

These are particularly severe as the absence of any common reference standard for assessing and identifying the "drivers" of psycho-social and socio-economic need makes it extremely difficult to start to develop shared in-depth multi-agency process pathway delivery models. Moreover it also precludes any attempt at establishing an equivalent epidemiological approach for well-being, and its potential links with social deprivation and other associated issues.

It is therefore somewhat unfortunate that such an embryo system was developed over a decade ago as a pilot project for the UK South East Thames Regional Health Authority but was effectively "stillborn" due to an NHS reorganisation. Its structure was developed from an in-depth analysis of a wide range of adult mental health needs assessments and subsequently tested on a further 300 plus set of assessments.

It broadly followed the same hierarchical approach used in ICD-10, which it was designed to supplement and support. Its top level consisted of three generic

NL - Legal	ALL aspects relating to an individual's position with regard to legislative constraints and issues arising therefrom
NE - Economic	ALL aspects relating to an individual's economic activity and financial situation
NS - Social	ALL aspects involved in recreational and social activities and relationships
NO - Occupational	ALL aspects involved in the major focus of an individual's normal daily activities
NR - Relationships	ALL aspects of close interpersonal relationships within a family or equivalent setting
NC - Contact	ALL aspects involved is enabling effective contact to be made and maintained with other appropriate individuals, society or its services
NV - Values	ALL aspects involved with prevailing personal, social and cultural value systems and standards
ND - Development	ALL aspects of the evolution or normal growth in the ability to survive and thrive both as an individual and also within society
NA - Accomodation	ALL aspects involved in ensuring that the individual has an appropriate environment within which to survive effectively
NH -Health	ALL needs associated with the improvement in and/or maintenance of an individual's personal health and safety that is subject to their own direct control

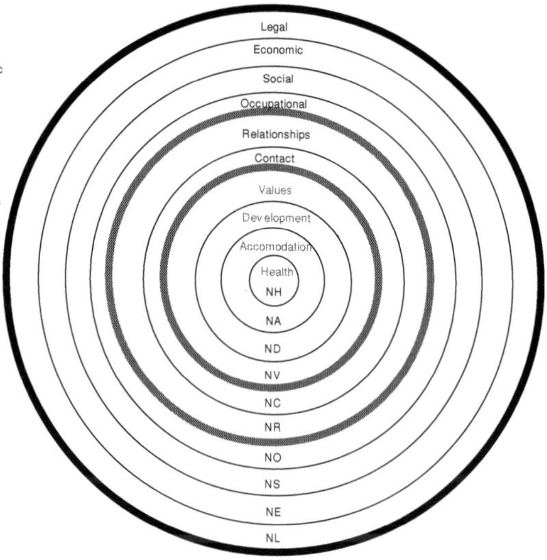

Fig. 4 Primary areas of need

domains that are nested, one within another, like "Babushka Russian Dolls". The basic self-needs of the individual sit at the core, surrounded by the direct interactive-needs related to immediate family/friendship ties, these are in turn set within the wider social needs of living within the bounds of civilised society (Fig. 4). Each of these ten primary areas of need are then expanded to define a more detailed set of categories (Fig. 5).

NR - RELATIONSHIPS

ALL aspects of close interpersonal relationships within a family or equivalent setting

NR1 - Basic	ALL needs resulting from failures in securing, building and maintaining effective relationships
NR2 - Partners	ALL needs resulting from failures, inadequacies or incompatibilities within a relationship
NR3 - Children	ALL needs arising out of difficulties or failures in coping with offspring through all phases of their development and their attitudes/behaviour
NR4 - Siblings	ALL needs arising out of all forms of relationship difficulties between children and their effects on the family
NR5 - Family	ALL needs resulting from dynamic and relationship imbalances, conflicts and communication issues within the overall relationship
NR6 - Parenting	ALL needs arising out of inadequate or dysfunctional child raising abilities/ skills, possibly resulting in bereavement or loss
NR7 - Abuse	ALL aspects of need resulting from violence, sexual, physical, emotionl abuse or ill-treatment
NR8 - Breakdown	ALL aspects of need resulting from either temporary, partial or complete dissolution of the relationship/s and its affects
NR0 - Other	ALL other Relationship-related needs that fall with the specified area, but which do notfall with the types of need defined above

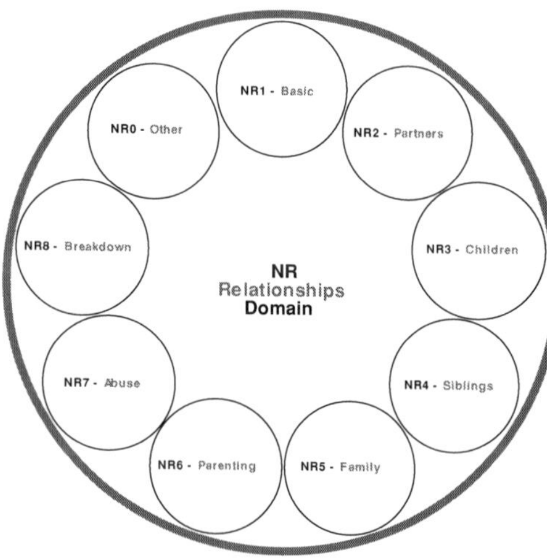

Fig. 5 Relationships—areas of need

As illustrated all definitions follow a set theory approach and encompass all needs within a specific domain area, and which cannot have its defined properties shared across its set boundary.

4 Complexity

Sadly in today's stress-ridden environment both multiple cases and the triggers for a range of psycho-social and socio-economic ills as well as dysfunctional relationships and likely disorders are to be found in a wide range of family situations.

The use of Genomaps gives an overview of family dynamics, which are often the cause of major tensions and breakdowns and can graphically illustrate clues to complex relationship issues, as well as significant social and economic problems (Fig. 6). In this example it is immediately evident each family member is not only under major stress but is also a contributor to family relationship problems; equally externally driven pressures act as further major stressors compounding and escalating the issues.

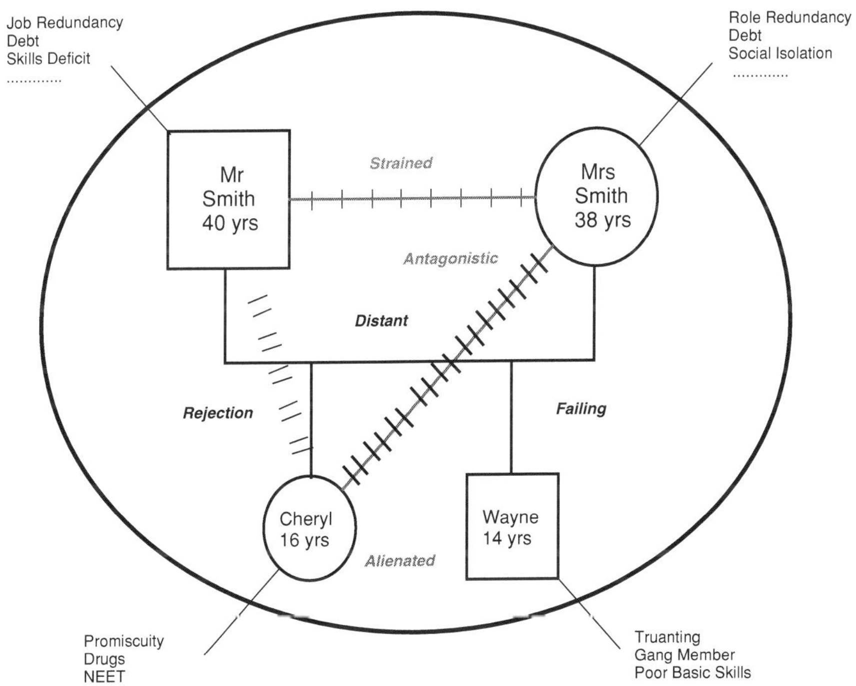

Fig. 6 Genomapping the roots of a family problem

Unfortunately this is not an unusual situation, which can only be resolved by forensically "unpicking" the individual component "diagnoses" and then setting about implementing pathways that each lead out of the morass. However such an approach depends on closely coordinated multi-agency, multi-disciplinary action, which is currently almost totally absent.

5 Coping Strategies

It is intriguing to note that the potentially vast amount of tacit knowledge of coping strategies adopted by so many people—past and present—has never really been tapped. Indeed this is encapsulated in an African Proverb [7]:

> "When an old man dies, a library burns down."

If action to collect, collate, analyze and map this as sets of self-care pathways for different situations, it would provide a valuable resource for the impaired and the care professions alike.

6 A "Virtual Library" System

This could be achieved by a further generic development of a concept proposed for the creation of a trans-European "virtual library" system for rare diseases [8], focusing initially on dysmelic conditions—resulting from both genetic and neurotoxin initiated damage to the fetus.

Its objective would be to develop an ever-evolving knowledge library acquiring tacit knowledge based on coping strategies adopted by interviewees to overcome life event problems (Fig. 7). Raw input would be processed, analyzed and content mapped using cognitive mapping techniques that would not only provide a sophisticated detailed summary but also act as an indexing mechanism for the stored source material [9].

As the system would be sourced and used trans-nationally, it must accommodate a range of cross-cultural and linguistic issues, which would involve incorporating a semantic ontology sub-system to act as a central multi-lingual reference thesaurus.

In view of the highly diverse nature of the latent audience a further key problem to be overcome is how to accommodate the many ways that individuals may perceive, search and then formulate their enquiry. In order to provide responses in an as readily intelligible fashion as possible, the aim is to allow the enquirer to define their preferred extract format profile.

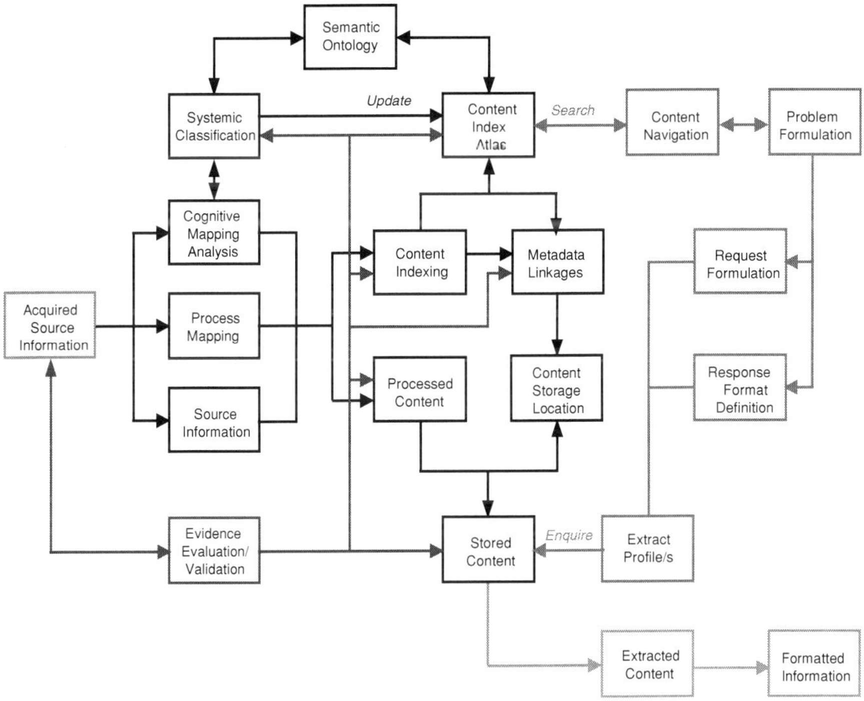

Fig. 7 Knowledge library system

7 Conclusion

The founding WHO definition of health as a balanced combination of well-being and healing spanning the physical, mental, and social domains as a potentially single entity has yet to be realized. Whilst the major stumbling blocks to this have been the complexity of the task and the resulting compartmentalizing into specialist "silos", it can potentially be overcome by developing a shared strategic model of service delivery based on the diagnostic led pathway approach successfully used in medicine.

References

1. Preamble to the Constitution of the World Health Organization as adopted by the International Health Conference, New York, 19–22 June 1946; signed on 22 July 1947 by the representatives of 61 States (Official Records of the World Health Organization, no. 2, p. 100); and entered into force on 7 April 1948.

2. Manning BRM. Coping with increasing infirmity: smarter smart support. Harrogate: Healthcare Computing; 2008.
3. Huppert FA, Baylis N, Keverne B, editors. The science of well-being. Oxford: Oxford University Press; 2005.
4. WHO ICD-10: international statistical classification of diseases and related health problems: tenth revision. 2nd ed. ISBN 92 4 154649 2 (vol. 1) (NLM classification: WB 15) 92 4 154653 0 (vol. 2) 92 4 154654 9 (vol. 3).
5. WHO ICF: International Classification of Functioning, Disability and Health. http://apps.who.int/classifications/icfbrowser/.
6. Paykel ES, Pruso BA, Uhlenhuth EH. Scaling of life events. Arch. Gen. Psychiatry 1971; 25:340.
7. The Last Word. The Times 2010 Apr 15.
8. EURORDIS. Report on the European Workshop on Centres of Expertise and Reference Networks for Rare Diseases, Prague, 12–13 Jul 2007. http://www.eurordis.org/IMG/pdf/EU_workshop_report_3.pdf
9. Manning BRM, Johnson M, Marsh A. Coping with clinical conditions in context: a European Knowledge Service. Proceedings of WorldComp 2009; 2009 Jul; Las Vegas.

Inverse Function Theory for Hearing Correction via the ABR in Memory of Swamy Laxminarayan

Koranan Limpaphayom and Robert W. Newcomb

Abstract The use of inverse function theory is discussed for hearing correction within the context of clinically measured ABR signals. Given an ABR for a modestly damaged auditory system it is shown how it can be corrected by a concatenation of two nonlinear systems to yield a type of hearing aid. Future extensions are mentioned.

1 Introduction

As Swamy was very interested in future visions for biomedical engineering, we present some visions via recent ideas towards the future of non-invasive correction of hearing and possibly other impairments. For concreteness we concentrate on the use of ABR signals which are now customarily used as a diagnostic tool [1].

Acoustic Brainstem Responses, ABR, and Otoacoustic Emissions, OAE, are two classes of signals presently used to evaluate hearing in situations where patients are unable to communicate with clinicians, such as for babies and animals [2]. These are signals which result from clicks or other known sounds applied to the ear. For the ABR they are measured as potentials on the scalp and for OAEs they are measured at the ear itself as reflected pressure waves. Since the signals are very weak they also require measurement in a sound controlled environment and quite sophisticated signal processing to eliminate background signals as may come

K. Limpaphayom and R. W. Newcomb (✉)
Microsystems Laboratory, University of Maryland, College Park, MD 20742, USA
e-mail: newcomb@eng.umd.edu

K. Limpaphayom
e-mail: koranan@eng.umd.edu

Commun Med Care Compunetics (2011) 1: 85–92
DOI: 10.1007/8754_2010_4

Published Online: 16 November 2010

from the heart beat or eye/body motion. But in any event through sufficient averaging the desired signals can be extracted and then used for clinical purposes [3]. Thus, actual responses can be compared to normal ones in order to determine damage to the audio signal paths involved in hearing.

Due to their clinical use there is an extensive literature on the ABR itself as well as means to measure and extract it [4]. The use of OAEs is somewhat older, and previously known as Kemp Echoes, with also an extensive literature. In both cases we have developed theoretical means, and transistor VLSI circuits to model the biological system that is processing the signals [5, 6].

Here we primarily discuss a means of correcting the ABR so that a damaged ABR can be restored to a more normal one. Many more details and hardware realizations can be found in the dissertation of the first of us [7] from which the figures created for that work are taken.

2 The ABR Signal

Figure 1 shows a typical normal ABR signal due to a click in one ear.

As shown there are five important peaks (and corresponding valleys) with their amplitude and time delays of interest. These are quite small being in the micro-volt range and usually embedded in milli-volt signals. They occur over a few milli-second time scale and show with a short delay from the exciting click. In Fig. 2 is shown the set-up for measuring the ABR for which three electrodes are typically used, one for the signal ground and two for differencing to remove background noise.

Figure 3 shows an ABR for a damaged ear.

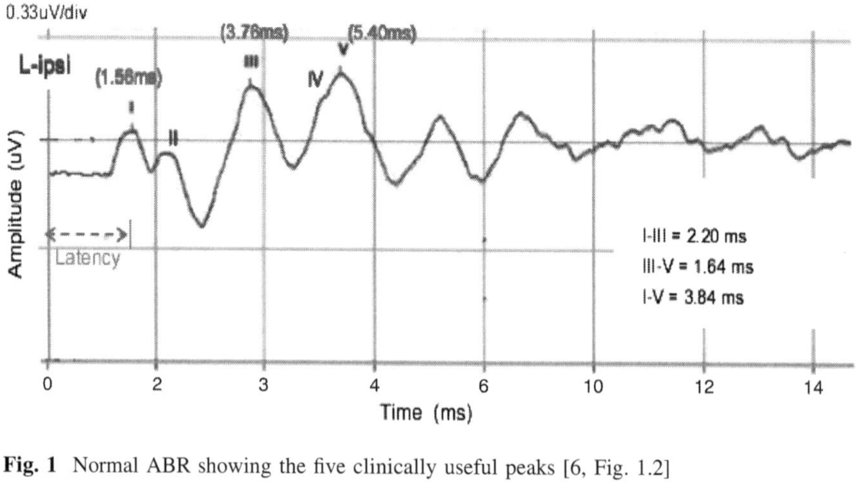

Fig. 1 Normal ABR showing the five clinically useful peaks [6, Fig. 1.2]

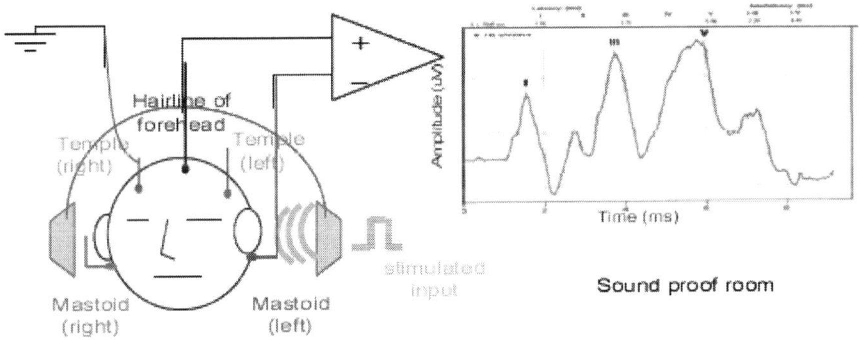

Fig. 2 Measurement scheme for ABR recording [6, Fig. 3.1]

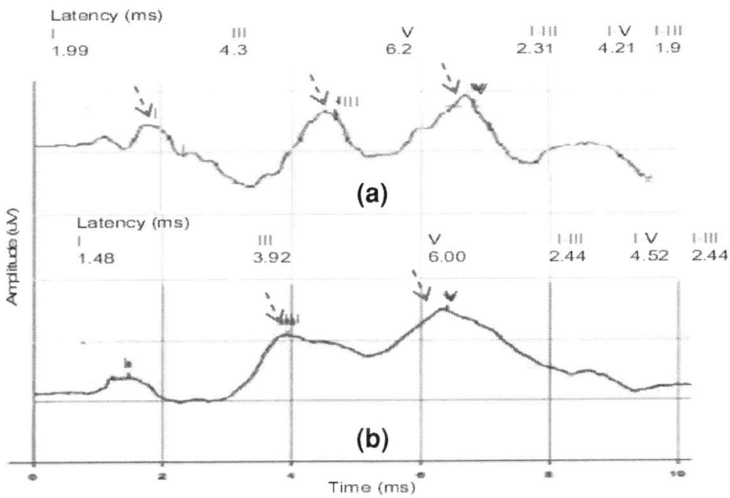

Fig. 3 Two abnormal ABRs **a** due to conductive hearing loss and **b** due to mild sensorineural hearing loss [6, Fig. 2.5]

As can be seen from Fig. 3 there is often enough differentiation in the peaks and valleys for a clinician to be able to tell that hearing loss has occurred without any vocal or other type input from the individual being evaluated. It is, though, natural to question what is it which determines the peaks and valleys and in a clinical sense what can go wrong to reshape them? For that purpose it is instructive to look at Fig. 4 where different audio pathways to the brain are indicated with an inference on how signals at each position may influence the signals detected on the scalp. To say the least, the there are myriad paths and almost definitely the measured signals contain numerous contributions from different paths. Even if one could be certain of where damage has occurred it would be a difficult task to carry out a corrective operation (though this is actually done in the case of cochlea implants!).

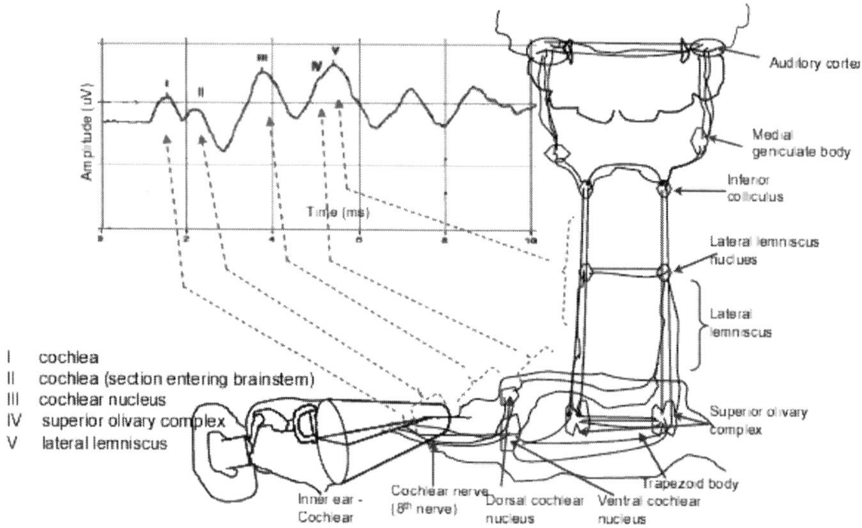

Fig. 4 Possible paths for ABR signals [6, Fig. 2.2]

By whatever means, the desire is to take a response as shown in Fig. 3 and turn it into one as shown in Fig. 1. Although this may not always be possible, in many cases it can be done. In fact the ABR is used in such a manner when implanting cochlear implants to be sure the implant is most effectively positioned. Our proposal is to do this via hardware worn in the same manner as present hearing aids such that no implantation is necessary. The key idea was expressed earlier [7] and is in the use of inverse function theory.

3 Inverse Systems for ABR Correction

Given a function $y = f(u)$, such as the ABR, the inverse function is given by solving for u to get a new function $g(y) = u$, in which case we write $f^{(-1)} = g$ and have $f(g(y)) = y$. That is $f(g(.)) = Id(.)$ where $Id(.)$ is the identity. In our case we have $y =$ ABR signal and $u =$ click in the ear. We have a measured ABR, y_{meas}, and assume that we know the desired ABR, y_{des}, both for the same click, u. Thus,

$$y_{mes} = f_{meas}(u) \tag{1}$$

$$y_{des} = f_{des}(u) \tag{2}$$

and we wish to insert a device in the ear to change y_{meas} into y_{des}. Since the body is the fixed part of the system, the aid signal will need to go through the aid before going through the fixed system; this means the function to be corrected, f_{meas} (.) should be inverted prior to its input. In other words, we wish to work with

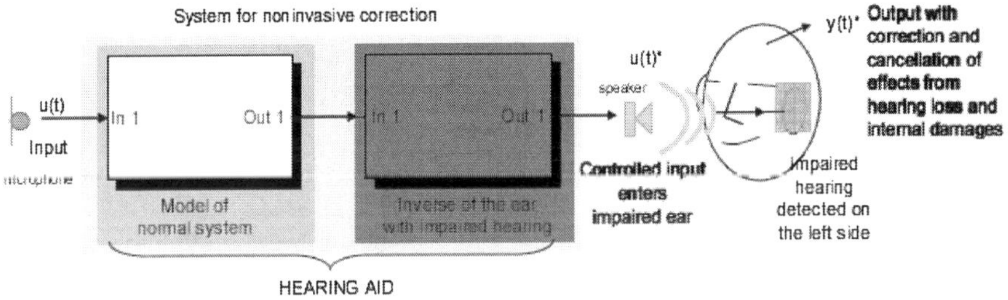

Fig. 5 Diagram illustrating use of the inverse system to correct the ABR [6, Fig. 4.2]

$$y_{des} = f_{meas}\left(f_{meas}^{(-1)}(y_{des})\right) \tag{3}$$

On substituting (2) into (3) we obtain

$$y_{des} = f_{meas}\left(f_{meas}^{(-1)}(f_{de}s(u))\right) \tag{4}$$

From this we see that we wish to make a system to insert into the ear which is described by

$$g(\cdot) = f_{meas}^{(-1)}(f_{des}(\cdot)) \tag{5}$$

That is, we wish to concatenate in front of the ear two systems, one for $f_{meas}^{(-1)}(.)$ and one for $f_{des}(.)$, combining them into one system described by the g(.) of Eq. 5. The ABR click signal u first enters the sub-system described by $y1 = f_{des}(u)$ and this sub-systems output, y1, is the input of the second sub-system described by $y2 = f_{meas}^{(-1)}(y1)$. The output of this second subsystem is fed into the ear whose ABR output is then $y = f_{meas}(y2) = f_{meas}(f_{meas}^{(-1)}(f_{des}(u)) = f_{meas}(g(u)) = f_{des}(u)$. In the end this gives the possibility of correcting the ABR from an abnormal one to a desired more normal one. Pictorially this mathematics is illustrated in Fig. 5.

This technique has been simulated on an abnormal ABR, obtained from Dr. Permsarp Isipradit of the Department of Otolaryngology, Chulalongkorn University, with the results shown in Fig. 6. As can be seen the results give an ABR much closer to the normal one. More details can be found in [6, 8].

4 Discussion

The use of an inverse system allows the cancellation of undesired responses and, as a consequence, can be used to correct some abnormal ABR and OEA recordings. If the system is linear taking the inverse amounts to division by a transfer function in the single input single output case. But if there is no ABR present this division is by zero pointing out that the method is not a cure all. However, in the case where damage is recoverable the use of this inverse system method could give

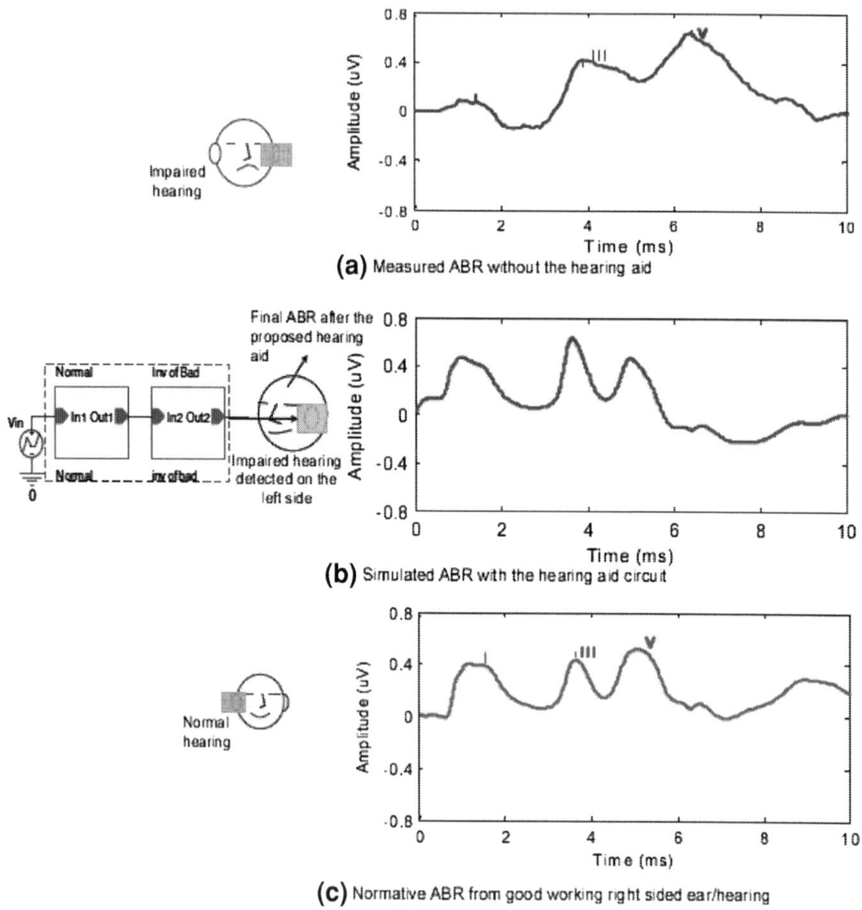

Fig. 6 An example of abnormal ABR correction [6, Fig. 5.13]

very favorable results, though as yet it has not been put to practice on actual human patients. Nevertheless, in [6] transistor Very Large Integrated Circuits are designed which could be fabricated to make hearing aids based upon the material developed here.

As seen by Fig. 4, where the auditory pathway is rather complicated and consists of a number of neural interfaces, the function determining the ABR is quite nonlinear. In fact large signals exciting the ear are necessary for the clinician to be able to have strong enough signals on the scalp for extracting an ABR. That being the case the mathematical determination of the inverse is not straightforward. Consequently, we have turned to the use of neural networks in describing the auditory system since neural networks are quite capable of capturing system nonlinearities. Again details are given in [6]. In some cases one may be able to linearize. Such has been successfully done in the case of OEAs [5] for which digital filters can serve as the description of the system for which the inverse replaces poles by zeros, and vice versa, in the z-transform domain.

Different clinics use different signals for measuring the ABR [9]. For instance we concentrated upon the use of a click for the input to the ear. But some clinics use speech sounds such as "da" What is important for the method discussed here is that the actual input giving the output is known since the output detected as the ABR does depend upon the input. Also in the choice of an input one would like signals which can serve as mathematical basis functions in order to adequately describe mathematically the class of systems under discussion. For the ear these are often sine signals with frequencies over the hearing band. But clicks can be seen to be similarly useful since their Fourier transform is frequency rich.

Normally the ABR is measured with one ear excited but since there are two ears the input could excite both in which case the input u becomes a 2-vector. One can also place many sensors on the scalp (in fact the 10–20 electrode system is an important standardized system) in which case the output y also becomes a vector. Consequently, the theory is appropriately developed for multi-input multi-output, MIMO, systems. This necessitates the inversion of nonlinear MIMO systems, for which again neural networks appear as among the most important available tools. Of course the mathematical theory is not limited to auditory systems and could be applied to sight, smell, taste, or even control of limbs.

Going beyond what is discussed above, we can turn the systems around and use the inverse system as the actual one. In the case of reciprocity this would mean that when we put ABR signals into the scalp we would obtain coming out of the ears the clicks which went into the ear to give them. Although reciprocity from the ear to the scalp probably does not hold, reciprocity from the scalp into neurons of the brain may. If so, we could externally make a system to take sound waves into ABR signals and then apply those ABR signals to the scalp so that a person with totally destroyed audio pathways may still be made able to hear. Since the scalp acts as a good insulator to electric fields, a system to do this would preferably work with magnetic fields. Once this is realized, one can conceive of using ABR type magnetic signals for control of many biological functions, including such things as epileptic fits and Parkinson's tremors.

References

1. Hall III JW. New handbook of auditory evoked responses. Boston: Pearson Education; 2007
2. Auditory brainstem response. Wikipedia.http://en.wikipedia.org/wiki/Auditory_brainstem_repsonse.
3. Silman S, Silverman CA. Auditory diagnosis: principle and applications. Academic Press; 1991.
4. Much beside [2] is available on the web. Some other clinically oriented sources are: http://www.hearingreview.com/issues/articles/2007-04_05.asp http://www.emedicine.medscape.com/article/836277-overview.

5. Sellami L, Newcomb RW. Ear-type analog and digital systems. In: Pandalai SG, editor. Recent research developments in circuits and systems. Trivandrum: Research Signpost; 1996. vol 1, p. 59–83.
6. Limpaphayom K. Microelectronic circuits for noninvasive ear type assistive devices. Doctoral dissertation, University of Maryland; 2009.
7. Gomez P, Rodellar V, Newcomb R. A PARCOR characterization of the ear for hearing aids. Proceedings of the IEEE; 1982; vol 70(12), p. 1464–6.
8. Limpaphayom K, Newcomb RW, Isipradit P. Ear type circuit and system simulating the auditory brainstem response for auditory disorder characterization. Proceedings of the 2009 IEEE/NIH Life Science Systems and Applications Workshop (LiSSA 2009); 2009; Bethesda. p. 66–6.
9. Jaspers HH. The ten-twenty electrode system of the International Federation. In: Deuschl G, Eisen A, editors. International Federation of Clinical Neurophysiology recommendations for the practice of clinical electroencephalography. Amsterdam: Elsevier; 1983. p. 3–10.

Emerging Use of Behavior Imaging for Autism and Beyond

Ronald Oberleitner, Uwe Reischl, Timothy Lacy, Matthew Goodwin and Josh S. Spitalnick

Abstract Commercially available "behavior imaging" technology is effectively assisting the diagnosis and management of children with neurodevelopmental disorders, including autism. This technology offers a unique way of capturing behavior data in natural environments on video clips, and is complemented by a comprehensive information storage and retrieval platform. Uses to date include providing families living in remote areas with improved access to health care services. The flexibility and versatility inherent in this new platform allow for future expansion, which may include a variety of wireless physiologic sensors. The benefits offered by this new technology have been recognized internationally and the use of this new technology may expand from behavioral healthcare and special education to fields such as psychotherapy, therapy supervision, and a host of research applications.

Keywords Autism · Autism technology · Behavioral health technology · Telehealth · Telemedicine · Telepsychiatry · Clinical supervision

1 Introduction

In response to the increasing demand by healthcare professionals internationally to include behavior information more systematically in the diagnosis and management of neurological disorders, academic and industry partners developed a new imaging technology which is now capable of providing comprehensive

R. Oberleitner (✉), U. Reischl, T. Lacy, M. Goodwin and J. S. Spitalnick
Boise, USA
e-mail: admin@talkautism.org

Commun Med Care Compunetics (2011) 1: 93–104
DOI: 10.1007/8754_2010_5

Published Online: 24 December 2010

documentation of relevant patient behaviors. Behavior Imaging®[1] now has the potential of transforming the healthcare industry in a manner similar to the impact observed when medical imaging was introduced (e.g., X-rays, MRIs, CT scans). "Behavior imaging" refers to the capture and secure sharing of patient behavior information using video and other digital electronic means. The behavior information is ideally collected within a patient's natural environment such as their home or a child's classroom, where it can be securely shared by patients and providers independent of geographic location. The available data can be utilized for diagnostic purposes, improving treatment protocols, training new healthcare professionals, and other purposes.

If any data ought ever be considered an essential clinical tool, it is video data, because it can capture important visual manifestations of symptoms associated with neurological disorders such as autism, Traumatic Brain Injury (TBI), Neuropsychiatric disorders, Alzheimer's and other conditions. This is important precisely because these conditions are defined, wholly or in part, by behavioral characteristics.

To be fully adopted, a behavior imaging technology platform must meet the security, privacy, and control requirements associated with the multifaceted legal landscape of healthcare industry.

2 Commercial Examples

Behavior Imaging systems currently available on the market feature novel video capture, clinical annotation tools, and a personal health record platform to capture, store, and securely share behavior data between and among caregivers and healthcare providers.

Behavior Capture™ This system was developed at the Georgia Institute of Technology's College of Computing. It consists of a novel video capture technology which can be used in a home or institutional environment and features a unique video buffering capability that documents relevant events that occur before, during, and after a behavior. This "going back in time" feature can provide insight into causes or triggers of certain behaviors (Fig. 1).

Behavior Connect™ This is a data and records management platform which allows users to organize, analyze, and share videos as well as other types of images such as X-rays and pertinent documents, among and between patients, caregivers, health providers, therapists, etc. It is a secure, HIPAA-compliant web-based application and complies with the healthcare industry's current health record technical standards. It facilitates the non-disruptive integration

[1] Behavior Imaging is a registered trademark of Caring Technologies, Inc. (aka Behavior Imaging Solutions) —Boise, ID, USA.

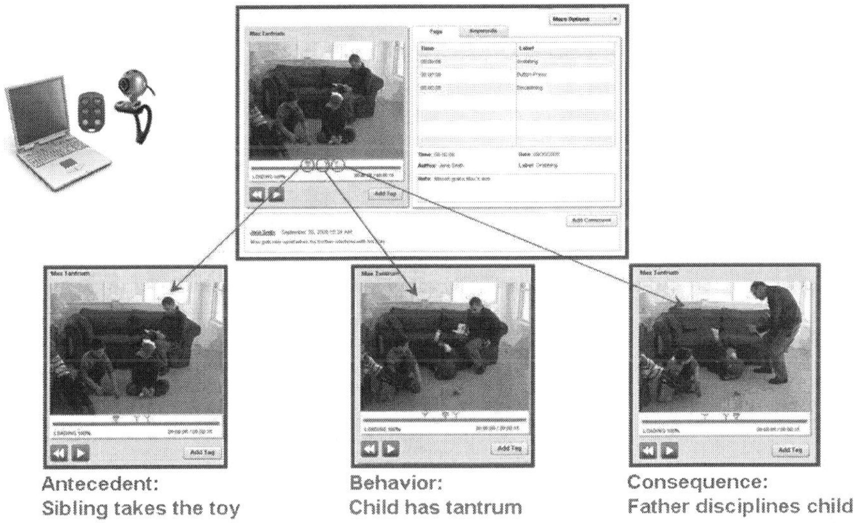

Fig. 1 With software applied to any personal computer, Behavior Capture™ enables a caregiver in a natural environment to capture a video clip of what happens before (antecedent), during (behavior), or after (consequence) with a small remote control device. That resultant video clip can tag aspects along the video clip that either the caregiver or the health professional can annotate for collaboration or data mining purposes [1]

Fig. 2 Behavior Connect™ allows clinicians to search file data of their patients (left), as well as uniquely tag and collaborate on any video or other data file (right)

with other database systems. The product also provides a proprietary fax transmission tool, special file uploader which can securely transmit large files even from low-bandwidth connections (as in rural areas), and a secure messaging system to enable electronic consultations. Field applications have shown that the technology is simple to use and can provide caregivers with important support during times of crisis or when asked to provide more contextual information (Fig. 2).

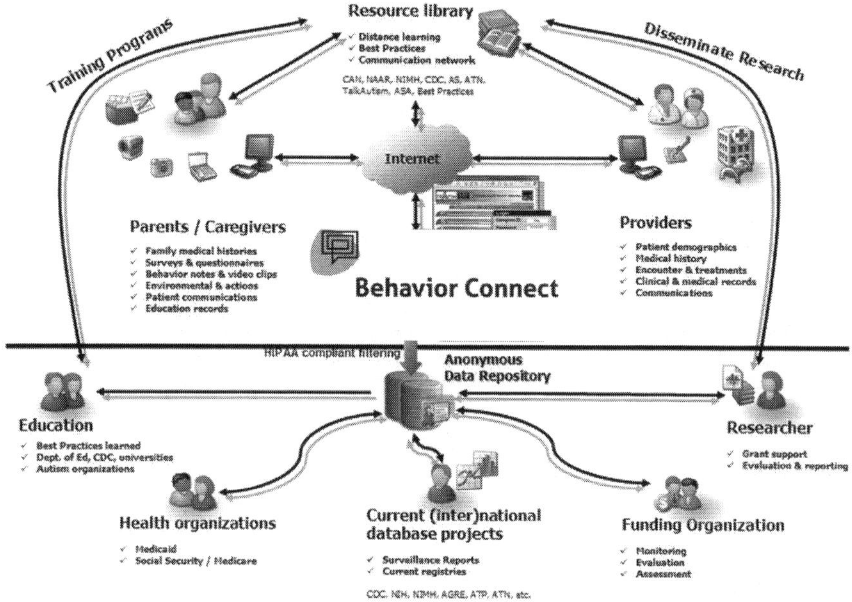

Fig. 3 Illustration shows how Behavior Connect^{TM} platform can connect caregivers from anywhere to their healthcare providers, and collect important data that can effectively treat the patient, but also provide researchers, health organizations, and others more understanding more about autism spectrum disorders [1]

Behavior imaging is available as a full solution to organizations or practitioners, or as a complementary plug-into existing health and special education databases (Fig. 3).

3 Use Cases

The efficacy of Behavior Imaging in promoting patient access to healthcare providers and improving data integrity in the clinical decision-making process has been evaluated by a number of leading academic institutions. The outcomes demonstrate high value in a variety of settings.

3.1 Health

3.1.1 Diagnostic Evaluation

Studies conducted at the Southwest Autism Regional Resource Center (SARRC), a leading autism center of excellence serving 11,000 individuals annually, showed

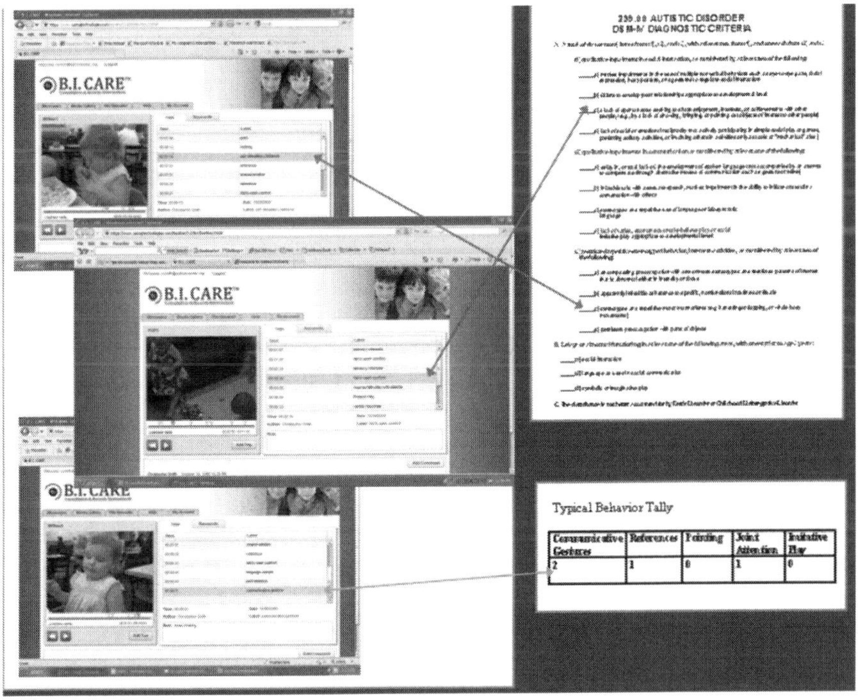

Fig. 4 Excerpt of a research poster illustrating Natural Observation Diagnostic Assessment. Illustrations show how three collected Behavior Image video clips taken from families with children with developmental delays from rural communities in Arizona, were transmitted remotely, received securely and later analyzed in Behavior Connect (formerly called "B.I. CARE", left side). Platform allows diagnosticians annotating to follow DSM IV criteria (top right), while providing tabulated scoring (bottom right)

that the use of Behavior Imaging accelerated the diagnosis of autism [2]. In addition, earlier work by the University of Medicine and Dentistry of New Jersey (UMDNJ) and Princeton Autism Technology (PAT) revealed that behavior imaging could provide valid diagnostic data directly from the home of the autism families [3]. These and other projects illustrate that this new technology platform can facilitate autism diagnoses, provide more accurate and more contextual information relevant to the management of the health condition, and serve as a resource by providing the ability to archive all relevant data in a child's electronic health record (Fig. 4).

3.1.2 Functional Behavior Analysis

Research has illustrated that if teachers use behavior capture to collect functional behavior data instead of conventional paper-and-pencil data collection, they experience 43% less errors when doing data collection for a functional behavior analysis, while providing collateral benefits. Researchers at Monash University in Melbourne, Australia documented the benefits offered by Behavior Imaging in providing a technology platform suitable for performing remote functional behavior analysis (FBA) and providing a data sharing system linking autism families living in the Australian outback with behavior specialists in Melbourne [4].

3.1.3 Remote Consultation

Families often have difficulty communicating their children's autistic behaviors to healthcare providers, and wait a long time to receive provider feedback. Many of these families are now using Behavior Imaging to share behavioral data and medical results in a more timely fashion. This strategy has reduced waiting times significantly when compared to parents who had to rely on in-person visits only [5]. In a telling practical example, data archived in a Behavior Imaging health record allowed one doctor to uncover a pattern of seasonal recurrences of behavioral issues in an autistic child which would not have been discovered any other way. In another example, many families displaced by Hurricane Katrina in 2005 were able to have their children with autism re-evaluated remotely to determine their behavioral health needs. This was possible because their health and behavioral data were saved with Behavior Imaging [6]. Rural families visiting SARRC in Phoenix, AZ have been shown to benefit from improved clinical outcomes when using Behavior Imaging to implement early intervention therapies. These therapies are now delivered and monitored from home [7]. Lastly, select US military dependents are now being helped by Applied Behavior Analysis (ABA) therapists who use Behavior Imaging to provide therapy which would have otherwise not been accessible.

3.1.4 Medication Management

As inferred in the previous section, behavior that occurs in the natural setting may have little relationship to behavior observed in the clinical setting. Behavior in the home can be captured, even "after the fact" using the time-buffering software, uploaded to a personal health record, and securely shared with specialists for the purpose of consultation, assessment, or follow-up of care. Patients who are difficult to transport, or for whom behavior in the office setting are of limited value, can receive initial or follow up care with potentially greater efficacy or acumen. Some preliminary support for this concept has been seen in studies that have used store-and-forward telemental health approaches for providing psychiatric and cross-cultural consultation to primary care providers in rural settings [8, 9].

One potential use for this approach is to use it for monitoring the progress of medication treatment. For example, families whose children are placed on medications would not need to return to the clinic for routine follow up appointments.

Rather, they can simply record behaviors at home and maintain a personal record. Child psychiatrists, child neurologists, or developmental pediatricians could follow the progress of patients using this web-based system with the hopes of providing care with a more accurate data set, make more appropriate medication changes, and avoid potentially negative medication side-effects.

3.2 Special Education

Education is currently the only accepted treatment for children with autism. Accepted treatment philosophies including early intervention has been shown to dramatically help children with autism, but experienced practitioners are in short supply and not easily accessible. Finding new ways to educate these children is a challenge.

3.2.1 Classroom Management

In a study funded by the National Institutes of Health (NIH), educators endorsed Behavior Imaging technology as a means of more effectively treating children with autism. Assessment surveys indicate that application of this type of platform in classroom settings is received favorably. Of the participating 29 educators from 11 different national clinical sites, 74% agreed that Behavior Imaging saved time and money by enabling them to easily capture on video what preceded a student's inappropriate behavior. 88% would use behavior capture to improve their teaching, and a majority of the participants reported that they will be able to serve more students than before. Application of this technology to staff training, student assessment, and supervision of students by their parents was reported as providing a significant benefit. "This would be tremendously helpful to our organization because we have 16 locations around the world and training and mentorship from central locations to the remote sites would be greatly enhanced with these capabilities", according to one participant (Fig. 5). Other participants commented that Behavior Imaging would address a critical need in rural schools, which often lack resident specialists [10].

3.2.2 Alternate Assessment

Special education students pose different challenges to be able to demonstrate their baseline skills, and progress they are making—due to many times not be able to take standardized tests. The Idaho State Department of Education (ISDE) is using Behavior Imaging to improve and simplify the process of creating, submitting and evaluating special education progress reports—and have this private student data available. This replaces the paper-based processes that have proven cumbersome

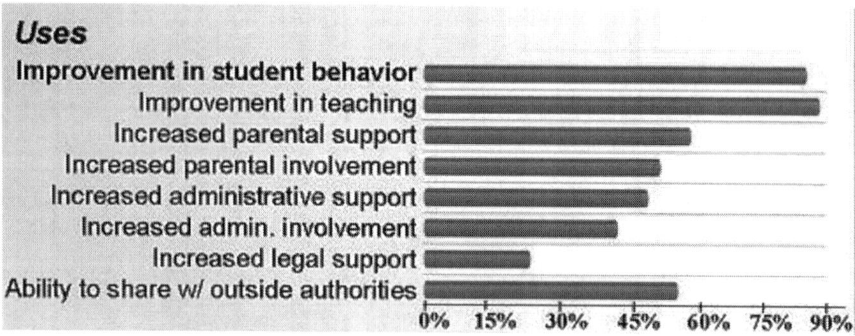

Fig. 5 Excerpt of a research poster illustrating perceived additional benefits of Behavior Imaging technology for special education classroom management

and expensive for teachers and administrators. Now, teachers in Idaho have captured and shared more than 70,000 video case studies, faxes or scans of homework or other reports electronically with the click of a mouse. Costs are estimated to be one half of doing this progress tracking through conventional 'paper' processes, while providing additional benefits.

3.3 Other Uses

The growing number of applications of the Behavior Imaging platform in the autism field has generated interest in this technology form other health fields, including:

3.3.1 Supervision of Psychotherapy in a Clinical Setting

Behavior Imaging systems would allow trainees to easily capture psychotherapy sessions (with patient consent) with the equipment they probably already have in the office, a computer and web-camera. After a session is conducted, key sections of the video can be "tagged" and notes added. The video can then be uploaded to a web-based "supervision record" where a supervisor provided with the appropriate passwords can log on, quickly scroll through the video, review key sections that the trainee tagged as important, and provide feedback notes both instantaneously and regardless of location. Supervisors no longer have to be on the premises to view confidential patient videos, and given that faculty are increasingly taking on more responsibilities, often at various sites, location is no longer an obstacle to timely supervision. This kind of system could improve the ability to provide direct/active feedback in a way that is less intrusive to patient and therapist, requires no additional equipment, is easy to use, easily shares the information with the supervisor, provides a platform for sharing information between trainee and supervisor, and creates a secure system for storing videos for teaching purposes and for documenting competencies.

The U.S. Air Force is developing a web-based supervision system using this type of approach for the entire Air Force Air Education & Training Command. Such a system would allow staff psychologists and psychiatrists to provide supervision to trainees within the same clinic, but also to provide supervision and mentoring to novice or newly graduated providers located in remote military locations anywhere in the world.

The U.S. Air Force is currently using Behavior Capture™ for a variation of the web-based concept just described. This is being done to capture the sessions of newly trained therapists who have learned a new intervention, virtual reality exposure (VRE) Therapy for the treatment of post-traumatic stress disorder (PTSD), a 5-year project funded by the U.S. Air Force and awarded to the fifth author and his organization, Virtually Better, Inc, based in Atlanta, Georgia. Having been trained by civilian psychologists who are among the leading experts in the fields of virtual reality and PTSD, the U.S. Air Force clinicians, in locations ranging from Anchorage, Alaska, to San Antonio, Texas, to Rheinland-Pfalz, Germany, receive continuity of training by meeting with these experts on a weekly basis, via telephone and videoconference consultations. The discussions of cases are driven by the videos that were captured and tagged via Behavior Capture™ and sent immediately to the civilian experts located in Atlanta, Georgia, and New York City.

3.3.2 Psychotherapy—Direct In-Session Applications

Behavior Imaging can be a valuable asset to the clinician in the context of psychotherapy. In its initial implementation phase, clinical psychologist Dr. Josh Spitalnick is using the Behavior Capture™ technology in a variety of unplanned and pre-planned ways, always with patient consent. The time-buffering capabilities inherent in this technology allow for the camera to be enabled and operational in "capture" mode to record unplanned critical moments in therapy (e.g., the patient has an insight several months in the making, the patient confronts a fear for the first time, or the therapist and patient experience an important therapeutic relational event). To do so, the clinician clicks the wireless remote, activating the software, which can "look back in time" up to 45 min to capture the powerful moment even though it has already passed. This video capture is then played back to the patient so he/she can experience it again and hear back the important revelation. This serves as a powerful experience in therapy whereby the patient gets to see him/herself in the same manner as the therapist and the world see them.

This use case of Behavior Imaging technology highlights the value of having video capture capability within psychotherapy treatment to allow the patient to hear his or her own words exactly as they were shared, and the additional capability of capturing time-buffered video. To capture unanticipated events even after they have occurred can provide a powerful and transformative moment in therapy. This technology also makes it possible to store this and any videos in the patient's personal health record (PHR), which is housed on a secure server and can easily be accessed, or store it in a clinical electronic health record for future review with the patient (if appropriate consent is given.

Clinicians can also use planned video-capturing in a variety of ways, including role-plays and capturing trauma narratives with patients engaged in imaginal exposure. Such videos can also be used, with patient consent, for training purposes by sharing them with other clinicians who can watch these video from more seasoned clinicians and witness techniques that foster a therapeutic relationship as well as observe the implementation of therapeutic interventions. For role-plays with patients, having video capture capabilities provides the clinician a valuable tool to share with the patient, by playing the role-play back, in session, so the patient can see how he/she comes across in the designed interaction. This allows the patient to more readily learn how they come across as effective or ineffective in a given scenario, and it provides the patient the opportunity to see the dynamic between two people, to learn what methods of communication are more or less effective.

3.3.3 Clinician Training

As many clinicians also serve as clinical mentors and supervisors to residents and graduate students, as is the case with several of the authors, one quickly begins to realize, one quickly begins to realize an added value to the capturing of unplanned and planned clinical interactions: a library of real training videos begins to take shape. Of course, anytime a clinician considers sharing patient data, written or digital, with anyone other than the patient, full informed consent must be addressed. The use of such technology can create a wonderful and rich set of training videos that are easy to view, share, and keep secure even when sharing. Behavior Imaging allows for a method of secure sharing of videos with trainees, experts, or supervisors, with the creator of the video being able to choose the level of security, from the recipient simply being able to view the video without downloading it, to being able to download, to being able to download it and provide digital feedback within the video. These and other uses have become an increasingly common practice among those who are comfortable introducing various forms of technology into the therapy process.

3.3.4 International Behavioral Research

A final exciting potential of this technology is in the collection of specific data regarding complex behaviors that can be compared across countries and across cultures. This data could be compared with and analyzed against other data sets such as environmental exposures, family histories, and genotype. This could potentially facilitate cross-cultural, large scale, neurobehavioral research that lends itself to culturally-informed evidence-based interventions. This system could facilitate nationally or internationally based multi-site research study in which medical, psychological, or behaviorally-oriented assessment or treatment research is conducted and require behavioral data to be evaluated, coded, and analyzed. Behavioral data could be related to the standardized implementation of the actual treatment protocol or the behavior of the actual research participant.

4 Future

With clinical and research benefit already, it is anticipated that the Behavior Connect technology platform will expand in several important ways. Complementary technologies such as the integration of wireless physiologic monitors that track blood pressure, EEG and EMG output, etc., can expand the type of information clinicians will have ready access—to make more informed decisions.

In the not-too-distant future, emerging technologies for unobtrusively measuring behavior wirelessly in natural environments [11] could be fruitfully combined with Behavior Connect. For instance, recent developments in ambulatory and non-contact autonomic nervous system monitors [12–15]), wireless accelerometers [16], and automated facial expression detection systems [17] could be used to explore physiological arousal [11], physical activity monitoring [18] and automated facial expression [17], respectively. Integrating these sensing modalities with time-locked, in situ video and audio collected by Behavior Connect would enable a rich, multi-level assessment of the environment and an individual's overt behavior and covert biology over time and across settings.

Furthermore, we anticipate in the coming years that real-time data mining and pattern recognition algorithms will enable automated extraction of important behavioral and developmental events that researchers, caregivers, and service providers can use to assess, understand, and support clinical populations from a distance. For example, imagine the value a clinician or researcher could glean if his 500 patients each having 100 behavior images captured can do. With today's information technology capabilities, one could use the Behavior Connect platform to pull up any examples where a patient articulated an intelligible word over years of data, or show if a patient ever had real eye-contact with another to demonstrate social abilities, or to pull up anytime his body made an unnatural behavior (based on norms that it can be compared to).

These may help make the review of complex behavior data as helpful, or even more helpful, than X-ray images or MRI images. Availability of Behavior Imaging offers great potentials for education, healthcare and research. The technology along with judicious clinical experience will bring never-before-available benefits to a wide variety of populations.

Acknowledgments The authors wish to recognize the late Dr. Swamy Laxminaryan for pioneering the conceptual framework of Behavior Imaging, which now is helping to achieve his dreams for improved health care through technology.

References

1. Oberleitner R, Wurtz R, Popovich M, Moncher T, Laxminarayan S, Reischl U. Health informatics: a roadmap for autism knowledge sharing. In: Bos L, Laxminarayan S, Marsh A, editors. Medical and Care Compunetics 2. Amsterdam: IOS Press; 2005.
2. Smith C, Ober-Reynolds S, Treulich K, McIntosh R, Melmed R. Naturalistic observation diagnostic assessment. International Meeting for Autism Research. 2009; Chicago, IL.
3. Oberleitner R, Laxminarayan S. Information technology and behavioral medicine: im-pact on autism treatment and research. In: Bos L, Laxminarayan S, Marsh A. editors. Medical and Care Compunetics. IOS Press; 2004.
4. Thomas J, Moore D, Anderson A. Functional assessment at a distance: an application of information technology in assessment and intervention. Applied Behavior Analysis International Conference. Oslo, Norway, August 7–9, 2009.
5. Oberleitner R, Elison-Bowers P, Harrington J, Hendren R, Kun L, Reischl U. Merging video technology with personal health records to facilitate diagnosis and treatment of autism. JD2H2 Conference, IEEE. Washington, DC, April 2006.
6. Reischl U, Oberleitner R, Simper P. Connecting autism families with emergency sup-port. Northwest Public Health. Fall, 2006.
7. Murphy P. Research center finds behavior imaging technology a powerful tool for treating children with autism. http://www.newswiretoday.com/news/63466/. Accessed Jan 18.
8. Hilty D, Yellowlees P, Cobb H, Bourgeois J, Neufeld J, Nesbitt T. Models of telepsychiatric consultation-liaison service to rural primary care. Psychosomatics. 2006;47:152–157.
9. Yellowlees P, Marks S, Hilty D, Shore J. Using e-health to enable culturally appropriate mental healthcare in rural areas. Telemed J e-Health. 2008;14:486–492.
10. Reischl U, Oberleitner R, Colby C, Choufrine A. Assessment of behavior imaging (B.I.) technology in the classroom. Association for Behavior Analysis International, Annual Conference. Phoenix, AZ, 2009.
11. Goodwin M, Velicer W, Intille S. Telemetric monitoring in the behavior sciences. Behav Res Methods. 2008;40(1):328–341.
12. Fletcher et al. iCalm: Wearable sensor and network architecture for wirelessly communicating and logging autonomic activity. IEEE Trans Inf Technol Biomed. 2010;14(2): 215–223 (March).
13. Poh M, McDuff D, Picard R. Non-contact, automated cardiac pulse measurements using video imaging and blind source separation. Optics Express. 2010;18(10): 10762–10774 (May 10).
14. Poh M, Swenson N, Picard R. Motion tolerant magnetic earring sensor and wireless earpiece for wearable photoplethysmography. IEEE Trans Inf Technol Biomed 2010;14(3): 786–794 (May).
15. Poh M, Swenson N, Picard R. A wearable sensor for unobtrusive, long-term assessment of electrodermal activity. IEEE Trans Biomed Eng. 2010;57(5):1243–1252 (May).
16. Tapia E, Intille S, Lopez L, Larson K. The design of a portable kit of wireless sensors for naturalistic data collection. In: Lecture Notes in Computer Science. Berlin/Heidelberg: Springer; 2006;3968.
17. El Kaliouby R, Picard R, Baron-Cohen S. Affective computing and autism. Ann N.Y. Acad Sci. 2006;1093:228–248.
18. Albinali F, Goodwin M, Intille S. Recognizing stereotypical motor movements in the laboratory and classroom: a case study with children on the autism spectrum. UbiComp; 2009. Sep 30–Oct 3. Orlando, Fl.

Factors Affecting Home Health Monitoring in a 1-Year On-Going Monitoring Study in Osaka

Toshiyo Tamura, Isao Mizukura and Yutaka Kimura

Abstract Home monitoring of blood pressure is beneficial in the management of hypertension. We have developed a new system for on-going home monitoring of physiological health markers, such as blood pressure (BP) and heart rate. The system was installed in 20 households in Osaka, Japan, with active participation by physicians. Of the 61 subjects in these households, 34 performed more than 100 BP measurements during the 1-year study period. Although the hypertensive subjects accurately and continually monitored their BP and other health parameters, the purported health benefits of monitoring appeared to be insufficient to motivate most of the healthy subjects to continue to make measurements. In this chapter, we discuss these results and the issues raised by this system. Although home BP-monitoring data are important in predicting cerebrovascular and cardiovascular disease, home monitoring systems and protocols need improvement, particularly in the area of compliance incentives.

Keywords Home healthcare · Blood pressure monitoring · Interface · Motivation · Compliance

T. Tamura (✉)
Graduate School of Engineering, Chiba University, Chiba, Japan
e-mail: tamurat@faculty.chiba-u.jp

I. Mizukura
Mitsubishi Engineering Co., Tokyo, Japan

Y. Kimura
Kansai Medical University, Moriguchi, Japan

Commun Med Care Compunetics (2011) 1: 105–113
DOI: 10.1007/8754_2010_9

Published Online: 16 November 2010

1 Introduction

On-going home monitoring of physiological health parameters can be beneficial to patients, providing data useful to the physician in the early diagnosis or prediction of disease. Blood pressure (BP) monitors for use in the home are widely available without a prescription, and home BP monitoring can be beneficial in the management of hypertension. It can motivate the hypertensive patient to maintain healthy habits, and BP variability is a good predictor of disease.

The American Heart Association recommends that all hypertensive persons obtain a home BP monitor and use it regularly to measure their BP at home. The Japanese Hypertension Society recommends that hypertensive patients monitor their BP twice per day [1]. The first measurement should be taken within 1 h of waking before taking any medications, and the second measurement should be taken before going to bed. In both cases, the subject should sit quietly for 1 or 2 min before taking the measurement.

According to Stergiou et al. [2], home BP monitoring provides more reliable and reproducible measurements than monitoring performed in a clinical setting or by an ambulatory BP monitor. In one study, the inclusion of home monitoring data led to the reclassification of clinical BP status in 54% of patients; 40% were reclassified downwards and 14% were reclassified upwards [3], demonstrating that home BP monitoring can play an important role in BP management.

In a long-term follow-up study of 5,211 subjects, 12.5% of the subjects exhibited white-coat hypertension (WCH), whereas only 10.8% exhibited home hypertension, and the risk of cardiovascular disease was higher for those with home hypertension than for those with WCH [4]. The relationship between WCH and normotension remains clear, but one study showed that WCH is a transitional condition leading to hypertension outside of medical settings and suggested that WCH may carry a poor cardiovascular prognosis [5].

Unfortunately, because precise and continual BP monitoring is rather cumbersome, patients must be strongly motivated if they are to continue such monitoring over a long period of time. However, the actual procedures for recording, storing, and displaying the data are simple and can be performed by subjects without any computer experience. In this study, we developed a new system for the monitoring of health parameters, including BP, and recorded data for 1 year. Here, we discuss our findings concerning the reproducibility of home BP monitoring and patient compliance.

2 Methods

2.1 System Configuration

Our system has three parts: a home sensor interface, a data-storage system, and a Web-based system (Fig. 1) [6, 7]. The data acquisition system connects to the health monitoring devices, which have an interface for data collection. The data

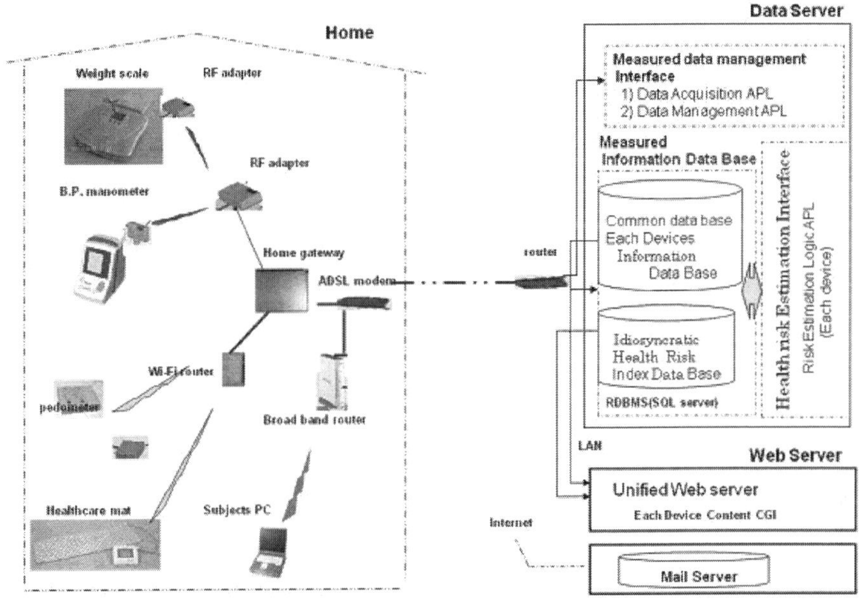

Fig. 1 The home health monitoring system used in this study

transmission and storage system connects to this interface and a central data server via the Internet. The Web server connects the Internet server to the subjects' personal computers and, with the subjects' permission, transmits information to healthcare professionals, including doctors, administrative dieticians, and physical trainers. A new feature of this system is its universal interface; subjects can use medical devices produced by different manufacturers because a unique identification number and standard protocol is available for each medical device.

All of the collected data are transferred to the central server over the Internet using a secure communication (Secure Sockets Layer) protocol. The central server creates a database for each subject and displays graphs of the data available on the Web server.

The system is installed in patients' homes to monitor health conditions and to serve as a gateway for data acquisition. Subjects can view their vital sign data on Web pages after entering a personal identification number and password. Options are available for displaying graphs of the data encompassing 1 day, 1 week, or 1 month at a time. For BP monitoring, the measurement time, systolic and diastolic BP, pulse pressure, and pulse rate (PR) can be stored and displayed. Duplicate data are omitted during the statistical analysis.

2.2 Study Population and Protocol

Sixty-one subjects (27 male and 34 female), including hospitalized persons, participated in and provided informed consent for this study. The average ages

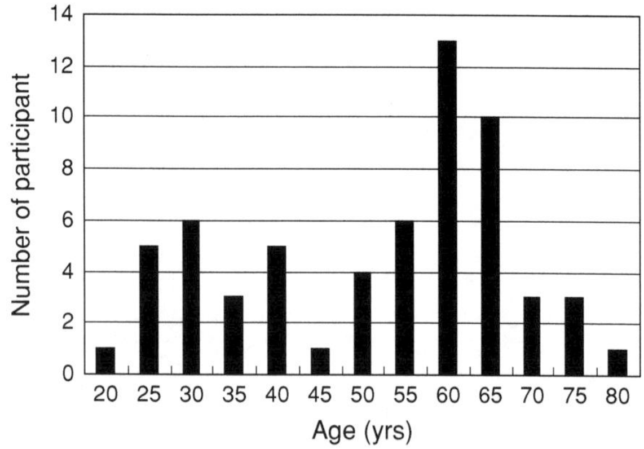

Fig. 2 Age distribution of study participants

(mean ± SD) of the males and females were 52.3 ± 15.52 and 46.0 ± 14.02 years, respectively (Fig. 2). The average systolic and diastolic clinical BP measurements for the entire study population were 138.0 ± 9.7 and 78.8 ± 7.0 mmHg, respectively. The BP distribution is shown in Table 1 according to the classes defined by the World Health Organization and the International Society of Hypertension, and frequency of signs and symptoms related to cardiovascular are shown in Table 2. Eight of the subjects were taking prescription drugs and were instructed to continue the drug regimens prescribed by their doctors.

The Ethics Committee of Kansai Medical University approved this study. Data were obtained for 1 year (25 January 2005 through 24 January 2006). Subjects monitored their BP using a commercially available BP monitor (CH-462E, Citizen Japan) that uses a semiautomatic oscillometric method and a standard arm cuff. The monitors were fitted with a digital communication tool for data communication. Where multiple users were to use a single monitor, up to four user-selection buttons were placed on the front of the monitor; each of the subjects was instructed to press his or her assigned button before taking measurements. All devices were validated and satisfied the criteria of the Japanese standard [8]. All data were automatically transmitted to the study data server via a home gateway.

The subjects were asked to measure their BP twice daily: once within 1 h after waking up in the morning and once before going to bed, based on the 2004 guidelines of the Japanese Hypertension Society [1].

Table 1 Clinical blood pressure statistics of study participants

BP status	Number of subjects
Original	2
Normal	7
Normal systolic value	15
Mild hypertension	8
Moderate hypertension	2
Severe hypertension	0
Total participants	34

Table 2 Signs and symptoms of study participants related to cardiovascular health

Symptom	Number of patients (duplicate)
Diabetes	3
Hypertension	8
Obesity	2
Dyslipidemia	10
Ischemic heart disease	10

Study subjects were trained by physicians and medical staff of Kansai Medical University for 4 weeks (27 December 2004 through 24 January 2005). The lengthy training was necessary because most subjects were not familiar with the use of medical apparatuses, and some were unable to read manuals. For BP monitoring, the medical staff taught the subjects the correct location for positioning of the sensor microphone above the brachial artery. Subjects were trained to measure their BP at home at predetermined times. The reliability and reproducibility of the BP measurements were poor at first but good at the end of the training period.

All subjects were given physical examinations at the clinic between 20 March 2005 and 10 April 2005. During the field study, the medical staff contacted the subjects every 2 weeks by e-mail or post.

2.3 Evaluation of Subject Compliance

We evaluated compliance by counting the number of continuing morning and evening measurements made by each subject. The target was 100 measurements made over a 50-day period.

3 Results

3.1 Subject Attrition

One subject withdrew from the study within 7 months, and five more withdrew within 9 months. Three of the withdrawers rejoined the study, so that six subjects completed the study in all.

3.2 Continual Monitoring Compliance

Of the 61 subjects completing the study, 15 males (average age 63.3 ± 6.9 years) and 19 females (average age 57.4 ± 11.0 years) made at least 100 BP measurements, meeting the compliance target (Table 3). In total, 58% of the subjects met

Table 3 Numbers of measurements made by subjects completing the study with at least 100 BP measurements

Parameter reported					
Subject ID	BP	Body weight	Steps	Sleep	Exercise
2001	662	614	330	332	188
2002	582	637	226	308	42
2005	406	571	321	336	20
2006	419	392	252		
2009	631	737	281	337	155
2010	611	468	203	351	46
2013	128	344	250	280	
2014	157	413	199	297	
2017	261	280	173		
2018	524	412	258		
2021	279	483	170	328	
2022	254	529		313	
2025	459	517	304	341	274
2026	441	485			167
2029	334	431	269	336	
2034	371	434	148		
2036	103				
2037	254	110	86	160	
2038	296	403			
2039	145	300			
2040	108	218	59	137	
2042	189	191	162	160	97
2045	352	313	283	287	
2046	168	247		248	
2049	446	503	293	273	12
2050	377	406	126		
2054	151	400	279		
2058	135	212	116		
2061	261	409	254	262	
2062	204	191	180		
2065	425	510	241	187	
2066	416	435	173		
2073	244	405	333	300	
2074	276	678	260		

the target. The data for those subjects with fewer than 100 sequential recordings were eliminated from the analysis.

3.3 Duplicate Measurements

For the 1-year monitoring period, we obtained 14,774 BP datasets. After verifying the data according to the measurement protocol, 1,706 of the BP datasets (11.5%)

Table 4 Morning and evening systolic and diastolic BP and PR

	Datasets obtained	Average systolic BP (mmHg)	Average diastolic BP (mmHg)	Average PR (beats/min)
Morning	327	129.56 ± 8.95	77.34 ± 7.52	61.63 ± 4.19
Evening	335	124.59 ± 10.12	69.94 ± 8.10	65.69 ± 5.55
Day	662	127.04 ± 9.87	73.60 ± 8.65	63.68 ± 5.33
Change[a]		5.52 ± 10.43	7.30 ± 10.29	–3.88 ± 5.78

[a] Morning value – evening value

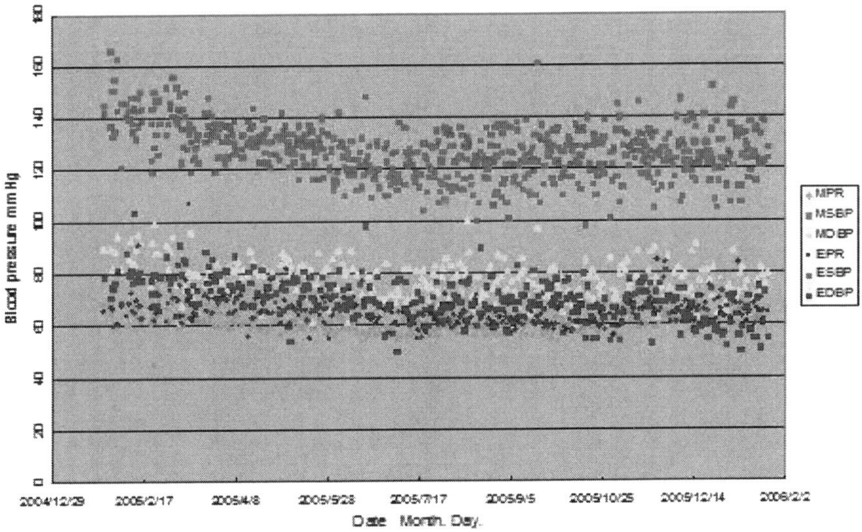

Fig. 3 Morning and evening BP and PR data for a typical subject

were found to be repeat measurements, where a repeat measurement is defined as one recorded after the one scheduled for that particular morning or evening had already been recorded.

The average time interval between repeat measurements was 3.75 ± 8.13 min. The average systolic and diastolic pressure differences between repeat measurements were 5.71 ± 12.07 and 4.25 ± 16.05 mmHg, respectively. The large standard error shows the reliability of BP monitoring. We assume that errors in positioning the microphone or pressure sensor or in applying the cuff were common.

The average morning and evening BP and PR measurements are shown in Table 4. BP values were higher in the morning than in the evening, whereas PR values were higher in the evening than in the morning (Fig. 3). The mean differences between the morning and evening systolic and diastolic pressures were 5.52 ± 1.43 and 7.30 ± 1.03 mmHg, respectively, and the mean PR difference was –3.88 ± 5.78 beats per min.

4 Discussion

4.1 Compliance

More than half of the study subjects (55.7%) in 20 households successfully monitored their BP continually over a 1-year period using a home-based data collection system. Withdrawals from the study were mostly attributable to a lack of motivation, particularly among the younger and healthier subjects, who were not very interested in health monitoring and found it inconvenient. Some subjects measured their BP once or twice per week but did not completely withdraw from the study. The main reason they stopped was the apparent ineffectiveness of healthcare monitoring. In our system, the data obtained from multiple devices was combined and analyzed together. Most of the subjects wore pedometers to count walking steps, one of the reported parameters. However, when walking did not control their BP, they ceased monitoring. Some subjects resumed monitoring after communications with study staff. We recommend that medical personnel contact home monitoring participants at least once every 3 months.

During the 4-week training period, participants measured their BP 759 times. Although home BP measurements are sometimes inaccurate, unreliable, and non-reproducible, our subjects were trained to set the microphone on the brachial artery and properly apply the cuff and were able to obtain reproducible data. Training was required to achieve good accuracy, reliability, and reproducibility. Acceptance of the home BP monitor was very important. During long-term monitoring, most subjects had a rough idea of their own BP. When they obtained unexpected values, they often repeated the measurement.

On-going BP monitoring is an effective method for preventing and predicting disease, but a highly accurate and reliable home BP monitor must be developed for home use. A more sophisticated but simpler to use home BP monitor and a simpler protocol are needed if on-going BP monitoring is to become widespread. In addition, data mining technology should be used to develop a prediction method.

4.2 Suggestions for Future Home Health Monitoring

Our findings suggest several factors that are important in home health monitoring:

First, personal identification must be simple. In our study, a single BP monitor and body weight scale were used by up to four people in each household. These devices were equipped with buttons to match each dataset with each subject in the household. Although the subjects were instructed to press their assigned identification buttons before taking measurements, errors in subject selection were an obvious problem. An error-proof personal identification method is needed for reliable monitoring.

Second, subjects benefit from an interactive "help" function. During our study, some subjects became interested in their health and in the principles of monitoring.

For example, some subjects asked the medical staff how to determine body fat percentages using the body weight scale. The subjects' questions and their answers were made available to all subjects on the study Web site.

Third, to maximize utilization efficiency, traffic patterns in the home should be taken into account when installing the monitoring devices in the home. If a single scale and BP monitor are to be used by all members of the household, they should be placed in a central bathroom used by all members of the household, each of whom would press his or her identification buttons before proceeding. However, this ideal situation is not possible in most Japanese homes due to lack of space; instead, the devices should be installed in a location central to all family members.

Fourth, interventions by medical personnel should be frequent and personal. Although we used e-mail for communications, face-to-face communication would be better. We suggest building a small community facility where subjects can visit at least every 2 weeks to talk with medical personnel about their health conditions.

Finally, incentives for home health monitoring should be initiated within the community. For example, subjects who successfully complete monitoring over a specified length of time could receive shopping coupons or a discount on local taxes. Reduction in health insurance costs might provide the funds for such programs.

Home monitoring of multiple physiological parameters predicts and prevents disease and should reduce the cost of health insurance. We suggest that improvements in monitoring systems and incentives should increase the number of participants in such programs.

References

1. The Japanese Society of Hypertension Guidelines for the Management of Hypertension (JSH 2004). Nippon Rinsho. 2005;63:952–8. (in Japanese)
2. Stergiou GS, Baibas NM, Gantzarou AP, Skeva II, Kalkana CB, Roussias LG, Mountokalakis TD. Reproducibility of home, ambulatory, and clinic BP: implications for the design of trials for the assessment of antihypertensive drug efficacy. Am J Hypertens. 2002;15:101–4.
3. Jai A, Krakoff LR. Effect of recorded home blood pressure measurements on the staging of hypertensive patients. Blood Press Monit. 2002;7(3):157–61.
4. Bobrie G, Genès N, Vaur L, Clerson P, Vaisse B, Mallion J-M, Chatellier G. Is "isolated home" hypertension as opposed to "isolated office" hypertension a sign of greater cardiovascular risk? Arch Intern Med. 2001;161:2205–11.
5. Ugajin T, Hozawa A, Ohkubo T, Asayama K, Kikuya M, Obara T, Metoki H, Hoshi H, Hashimoto J, Totsune K, Satoh H, Tsuji I, Imai Y. White-coat hypertension as a risk factor for the development of home hypertension: The Ohasama Study. Arch Intern Med. 2005;165: 1541–6.
6. Mizukra I, Tamura T, Kimura Y, Yu W. New application of IEEE 11073 to home health Care. Open Med Inform J. 2009;3:32–41.
7. Tamura T, Mizukura I, Kimura Y. Designing pervasive healthcare applications in the home. In: Coronato A, DePietro G, editors Pervasive and Smart Technologies for Healthcare. USA: Medical Information Science Reference; 2010. p. 282–94.
8. Japanese Industrial Standards Committee Cited 5 May 2010. Available from: http://www.jisc.go.jp/

Services-Based Systems Architecture for Modeling the Whole Cell: A Distributed Collaborative Engineering Systems Approach

V. A. Shiva Ayyadurai

Abstract Modeling the whole cell is a goal of modern systems biology. Current approaches are neither scalable nor flexible to model complex cellular functions. They do not support collaborative development, are *monolithic* and, take a primarily manual approach of combining each biological pathway model's software source code to build one large monolithic model that executes on a single computer. What is needed is a distributed collaborative engineering systems approach that offers massive scalability and flexibility, treating each part as a services-based component, potentially delivered by multiple suppliers, that can be dynamically integrated in real-time. A requirements specification for such a services-based architecture is presented. This specification is used to develop *CytoSolve*, a working prototype that implements the services-based architecture enabling dynamic and collaborative integration of an ensemble of biological pathway models, that may be developed and maintained by teams distributed globally. This architecture computes solutions in a parallel manner while offering ease of maintenance of the integrated model. The individual biological pathway models can be represented in SBML, CellML or in any number of formats. The EGFR model of Kholodenko with known solutions is first tested within the CytoSolve framework to prove it viability. Success of the EGFR test is followed with the development of an integrative model of interferon (IFN) response to virus infection using the CytoSolve platform. The resulting integrated model of IFN yields accurate results based on comparison with

V. A. S. Ayyadurai (✉)
Department of Biological Engineering, Massachusetts Institute of Technology (M.I.T.), 77 Massachusetts Avenue, Cambridge, MA 02139, USA
e-mail: vashiva@mit.edu
URL: www.vashiva.com

V. A. S. Ayyadurai
Systems Biology Research Group, International Center for Integrative Systems (I.C.I.S.), 701 Concord Avenue, Cambridge, MA 02138, USA

Commun Med Care Compunetics (2011) 1: 115–168
DOI: 10.1007/8754_2010_1

Published Online: 16 November 2010

previously published in vitro and in vivo studies. A open web-based environment for collaborative testing and continued development is now underway and available on www.cytosolve.com. As more biological pathway models develop in a disparate and decentralized manner, this architecture offers a unique platform for collaborative systems biology, to build large-scale integrative models of cellular function, and eventually one day model the whole cell.

Keywords Complex systems · Systems biology · Distributed computing · Systems architecture · Whole cell modeling · Distributed Collaborative Engineering (DCE) · Molecular pathways · Biological networks · Collaboratory

1 Background

A grand challenge of systems biology is to *model* the whole cell. A model for the purpose of this discussion is defined to be a mathematical representation along with its implementation in software including any input data and documentation. A cell consists of a set of organelles. These organelles interact through the medium of molecular interactions to provide *cellular functions* such as protein synthesis, metabolism, apoptosis, or motility. Systems biology aims to develop a model of the cell by connecting the biochemical kinetics of these interactions at the molecular mechanistic level to derive the quantitative descriptions of higher level cellular functions [1–5].

1.1 The Complexity of Biology

Biology is a field based on experiments, not first principles (ab initio) such as physics or engineering. It is fundamentally an experimental science. Biologists do many experiments to understand genes, proteins, protein–protein interactions. One example of perhaps the largest experiments in biology is the Human Genome Project (HGP) begun in 1990 and completed in 2005. This effort resulted in the discovery of only 20,000–25,000 genes, far less than what was originally theorized [6]. More interesting is the discovery that this number of genes is in the same realm as that of the nematode *Caenorhabditis elegans* which has approximately 19,000 genes [7]. More recently, the genome of the starlet sea anemone—*Nematostella vectensis*, a delicate, few-inch-long animal in the form of a transparent, multi-tentacled tube - was sequenced and found to have 18,000 genes [8].

Human and a nematode (or sea anemone) have a similar number of genes, but a great difference in complexity of function as whole organisms. This contradiction has led scientists to conclude that perhaps the number of genes in the genome is not connected with the complexity of the organism. Much of an organism's complexity can be ascribed to regulation of existing genes by other substances (such as proteins) rather than to novel genes [8]. The types and kinds of molecular interactions across the nucleus, cytoplasm and organelles, beyond the number of genes in the nucleus, may be the critical element in determining the difference between human and nematode, for example. This reasoning has led to an even greater activity to understand the structure of proteins (e.g. the product of genes) and protein–protein interactions.

1.2 Proteins and Protein–Protein Interactions

As of the writing of this document, approximately 30,000 proteins have been documented across various publications world wide [9]. Thousands of research teams across the world have contributed to these discoveries. Discovering the structure of just one protein is a difficult experimental effort. Such efforts are highly domain specific and one research team, for example, by itself may focus on understanding a small set of genes or the structure of a set of specific proteins or the interactions between certain types of proteins. The protein structures are determined from experiments using x-ray crystallography [10]. While new proteins are discovered each day, the crystal structure of most proteins is not known. In many cases, understanding just *one* protein structure requires the effort of not just this one laboratory's research team but the effort of multiple research teams spread across the world. New databases such as: HPRD, OMIM, PDB, Entrez Gene, HGNC, Swiss-Prot, GenProt are becoming repositories for storing protein structure as well as protein–protein interactions.

Approximately 40,000 protein–protein interactions are documented today and continue to grow. Each day, new protein–protein interactions are found. In addition, each day, updates of knowledge are made to existing protein–protein interactions. Experiments are used to derive such protein–protein interactions since these interactions cannot be derived from first principles. Moreover, sometimes, different research teams may get differing results for the same pair of protein interactions.

1.3 Biological Pathways

Biological pathways are networks of protein–protein interactions. Single protein–protein interactions can be combined to build biological pathways. Biologists, in addition to understanding the nature and function of genes, proteins and protein–protein interactions, also perform experiments to discover biological pathways. Today, approximately 60,000 biological pathway diagrams are recorded across a variety of databases including KEGG, Science STKE, Nature PID, BioPax, BioCarta and others. New biological pathways are being discovered each day.

A biological pathway consists of two elements: (1) molecules (also known as molecular species or species) and (2) interactions among those molecules. A biological pathway is visually represented using a "ball and stick" diagram as shown in Fig. 1. Each "ball" denoted by circles, rectangles or other geometric shapes represent the individual molecules. Each "stick" denoted by arrows or lines represents the molecular interaction. Because experiments not first principles determine the description of a biological pathway (e.g. which molecules will interact with another), biological pathways are constantly changing as new experiments reveal either changes in molecular species or nature of molecular interactions.

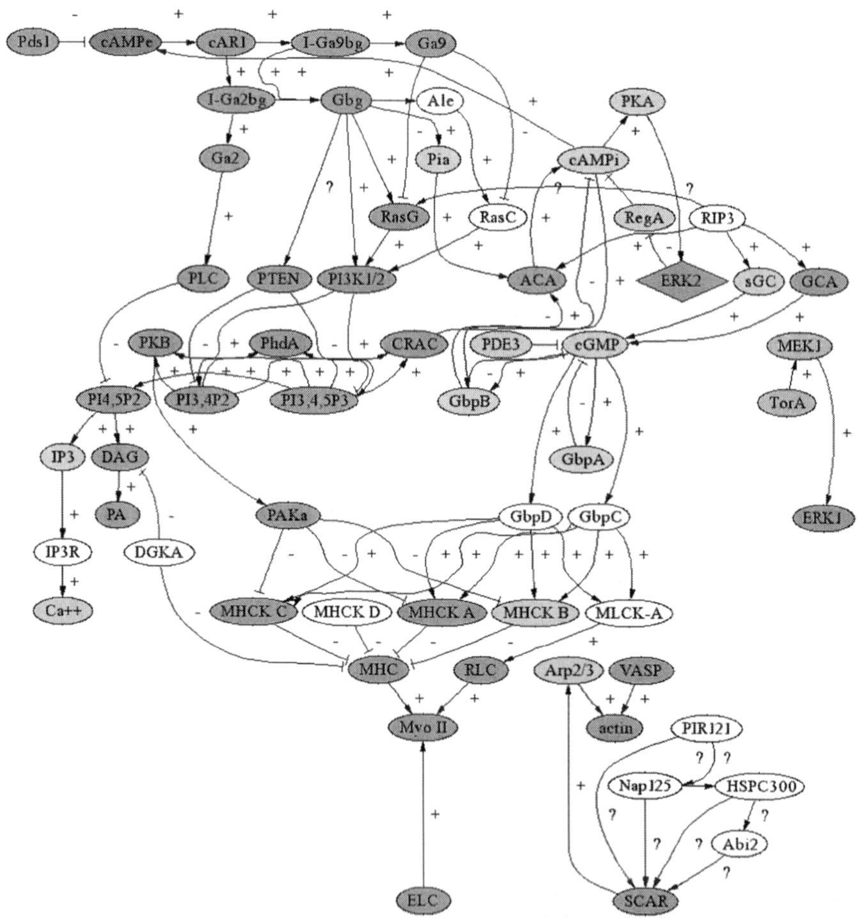

Fig. 1 Example of a biological pathway diagram containing many molecular species denoted by the geometric shapes of circles, ellipses and diamonds along with the multiple molecular interactions denoted by lines and arrows [86]

The biological pathway in Fig. 1 is the result of aggregating knowledge from nearly 35 different published articles. On average each biological pathway consists of approximately 5–15 molecular species and approximately 10–25 molecular interactions. The development of just this one biological pathway requires the integrated effort of multiple laboratories, spread across the world to construct and maintain. Elements of the biological pathway, the number of molecular species and the types of interactions, are subject to change based on new experimental results.

In summary, the development of each and every biological pathway is a highly collaborative effort, requiring the aggregation and integration of numerous experimental results, derived from the fundamental understanding of genes, proteins, and protein–protein interactions. Moreover, such a development effort, as will be discussed in forthcoming sections, is not linear, but cyclic, involving constant updates and refinements to the biological pathway, based on new experiments.

1.4 Systems Biology

Systems biology is a new field; however, building systems-level understanding of biology is not a new phenomenon. Over 6,000 years ago, many traditional systems of medicine including Siddha, Unani, Ayurveda and Traditional Chinese Medicine (TCM) proposed systems approaches to describing the whole human physiome [11, 12]. During modern times, starting in 1930s, with the concept of homeostasis [13] and biological cybernetics [14] attempts were made to understand biology from a systems level using the modern language of physics and control systems theory.

The discovery of the structure of DNA in 1953 [15] combined with recent high-throughput measurement techniques for imaging and quantifying molecular level interactions has enabled a completely new field of biology: systems biology. Systems biology aims to develop system-level understanding by connecting knowledge at the molecular level to higher level biological functions [16]. Such a goal was not possible before. Previous attempts at system-level approaches to biology, whether ancient or modern, were primarily focused on the description and analysis of biological systems, limited to the physiological level. Since these approaches had little to no knowledge of how molecular interactions were linked to biological functions, a *systems biology* of connecting molecular interactions to biological functions was not previously possible. Systems biology, therefore, is a new field of biology as it offers the opportunity, as never before in human history, to link the behaviors of molecules to the characteristics of biological systems. This new field will enable us to eventually describe cells, tissues, organs and human beings within a consistent framework governed by the basic principles of physics [17]. Systems biologists aim to link molecular-level interactions with cellular-level functions through quantitative modeling.

1.5 Biological Pathway Models

Over the past decade, new measurement techniques are enabling biologists to quantify the molecular concentrations and rates of molecular interactions within biological pathways. Such techniques are being used to transform diagrammatic representations of biological pathways to build biological pathway models.

Systems biologists convert "ball and stick" diagrams to *biological pathway models*, mathematical representations expressed in software program code along with the inputs and data needed to run the model. Figure 2 illustrates the high-level process by which a biological pathway is converted to a biological pathway model. There are approximately 300 published biological pathway models today.

The software program source code of these models may be represented in different formats: MATLAB, C++, C, SBML, CellML, etc. Different mathematical representations including ordinary differential equations (ODE's), Boolean Networks, Stochastic approaches, analytical functions, etc. are used to specify these models. The internal parameters for these models, such as kinetic rate constants, for example, are also determined through experimentation. Maintaining just one biological pathway model is a complicated task since the biological pathway models and the internal parameters are constantly updated based on changes to the biological pathway diagrams (e.g. based on new experiments). There are different emerging repositories for hosting biological pathway models including BioModels.Net and CellML.Org [18, 19].

From these repositories, one can download a biological pathway models and execute them on a local computer. Figure 2 illustrates the transformation of the biological pathway diagram, on the left hand side, to a *black box* biological pathway model, on right hand side. This black box has inputs, being the species concentrations of the molecular species at time $t = n$, and outputs being the

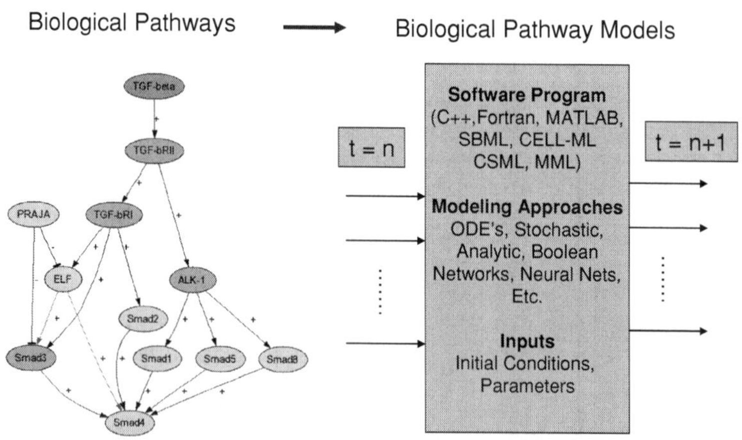

Fig. 2 Systems biologists work to convert biological pathway diagrams to biological pathway models

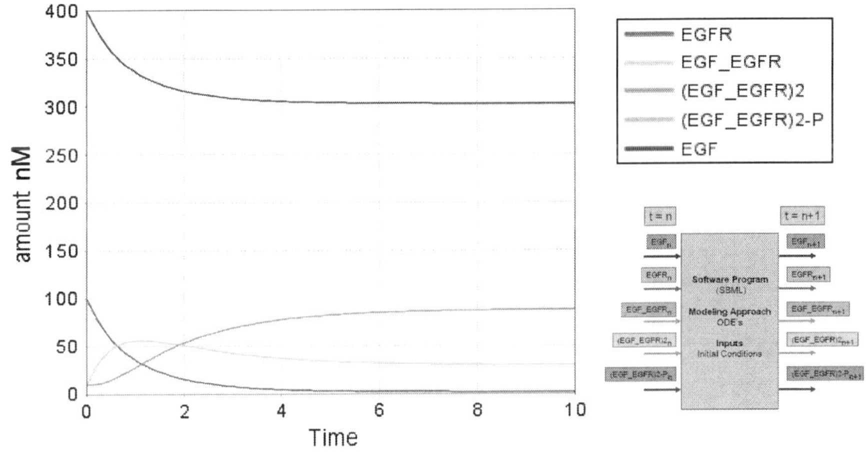

Fig. 3 Results from the execution of the biological pathway model

species concentrations of the molecular species at time $t = n + 1$. The internals of the black box contain the software code and the mathematical representation. In this case, the mathematical representation is an ODE and the software code is in SBML. However, the mathematical representation and the software code could be in any format as described earlier. Execution of this biological pathway model will yield results as shown in Fig. 3.

The results in Fig. 3 are the time varying changes in species concentrations. Along the x-axis is time and along the y-axis is the species concentration in nM (nano-molar). Such a biological pathway model serves to provide a quantitative and predictive capability to describe the interactions of five molecular species. Each biological pathway model is treated as a black box, having a defined set of input species and the same defined set of outputs, the values of which are the species concentrations. The construction and validation of such biological pathway models remains a tedious and time-consuming process mainly due to the experimental effort required to determine the many internal parameter values.

Most biological pathway models are developed, used for a single application by a single developer, and then forgotten. One can only imagine how many biological pathway models, for example built in MATLAB, and never published are located in some file folder, in some unknown computer, developed by some graduate student. Therefore, considering all the hard work that goes into developing a model, it is rather surprising that so little attention is paid to the presentation and conservation of existing models [20].

Systems biologists are attempting to create reusable components of biological pathway models by constructing online repositories to offer an archive and curated repository. In addition to supporting the conversion of the diagrammatic representation of biological pathways to biological pathway models, systems biologists also aim to build *larger* biological pathway models by integrating *smaller* biological pathway models. The purpose of such effort is to gain new insights on cellular functions not possible from experimental research. Currently, there are approximately 5–10 such integrated biological pathway models. There are three main reasons why there are such a low number of integrated biological pathway models.

First, the individual biological pathway models are in different formats. Second, understanding any one model requires a great deal of domain specific knowledge and expertise. Third, the primary method of integration involves merging the source codes of each biological pathway model into one large source code. Because of these reasons, it is very time-consuming and expensive to integrate biological pathway models. Maintenance of the resulting integrated model is also very difficult since the integrated model can become invalid as it has a "half-life." New proteins, protein–protein interactions and new parameters (e.g. rate constants) in any one of the individual biological pathway models are being discovered and/or updated constantly. Integrating models can also become especially difficult, within the current method, if the source code for any one particular biological pathway model is not publicly available. This would require one to recode that entire model's source code from scratch.

2 Motivation

Figure 4 illustrates the path to modeling the whole cell and summarizes the concept from the previous sections.

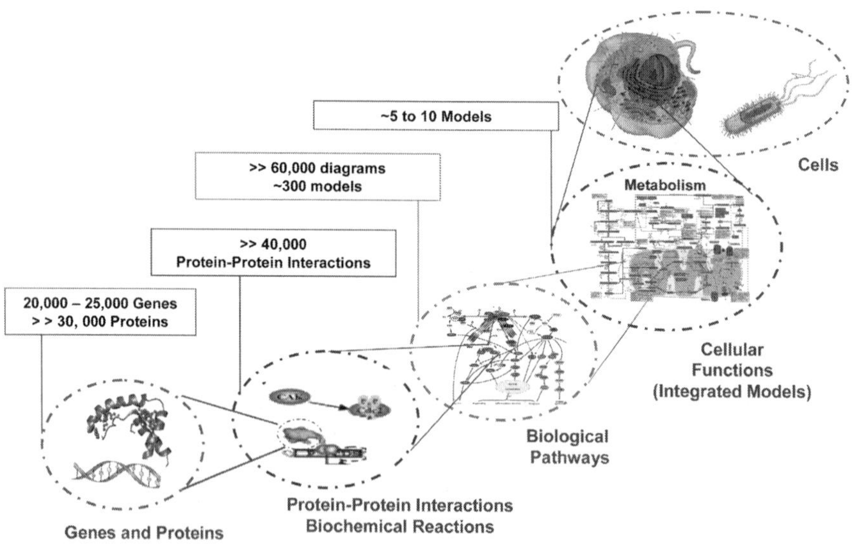

Fig. 4 Summary of the development path towards whole cell modeling

This figure presents four major steps to modeling the cell. First, the understanding of genes and proteins are used to build an understanding of protein–protein interactions. Second, these protein–protein interactions are networked to create biological pathways. Third, the integration of biological pathways serves to describe cellular functions. Fourth, and finally, the integration of cellular functions serves to model the whole cell.

Currently, as shown in Fig. 4, there are only 5–10 such integrated models of cellular function. It is expected that the number of biological pathway models, currently numbering approximately 300, will grow; however, the step in developing larger integrated models is severely limited by the time consuming and expensive effort, for the three reasons highlighted earlier. The discussion below provides greater insights on the efforts needed to build biological pathway models.

2.1 Development of Biological Pathway Models

The process to create and maintain a particular biological pathway model is an iterative process of manipulating, measuring, mining and modeling as shown in Fig. 5. The two major areas are experimentation and modeling. Systematic experiments involve manipulation and measurement. Manipulation involves modifying an existing biological system. Measurement involves collection of data from that manipulated biological system. Quantitative modeling involves both mining and modeling. Mining enables the identification of underlying relationships in large datasets.

These relationships can be used to create predictive mathematical models. Biologists, working in highly domain specific areas, perform systematic experiments, using a range of advanced measurement devices quantify the molecular

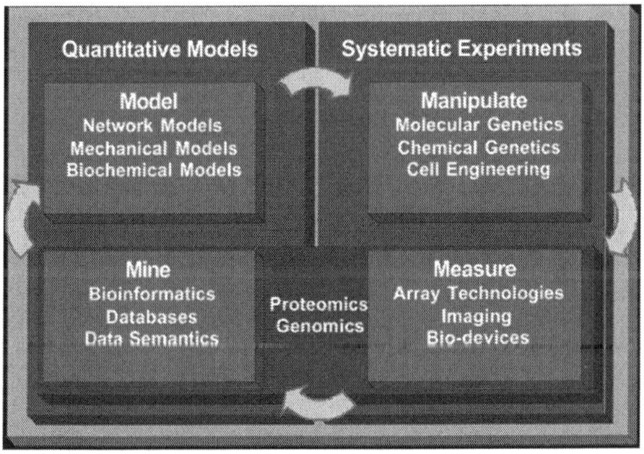

Fig. 5 The four M's of systems biology [87]

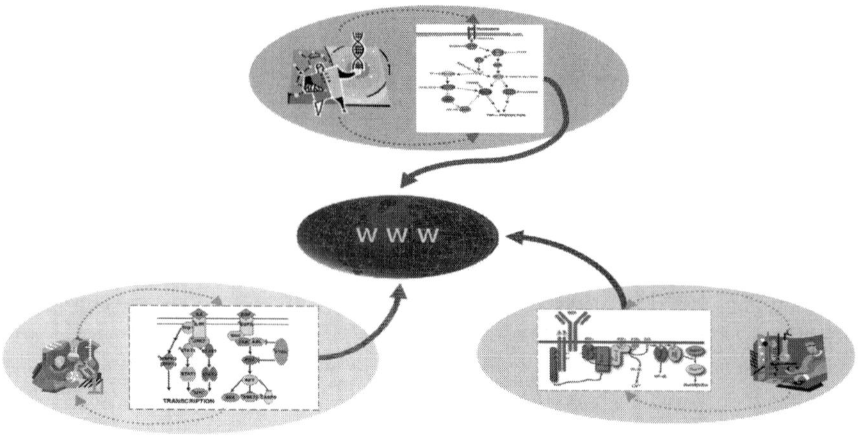

Fig. 6 Scenario of three research teams performing systematic experiments to produce biological pathway models, which are published and Complexity of Integrating Multiple Biological Pathway Models

concentrations and dynamics of molecular interactions. Data mining and modeling efforts are used to refine their conclusions. Biological pathway models are developed and refined through this constant and arduous iterative process. The World Wide Web (WWW) offers a vehicle for scientists to more easily share and publish their biological pathway models. There are thousands of such biological pathway models being published and refined each day by teams of biologists worldwide. Figure 6, for example, illustrates three different research teams, spread across the globe, performing the iterative process of systematic experiments and modeling to produce biological pathways, which are made available and published over the WWW.

Given the decentralized nature of these efforts, the source code of any one biological pathway model may be written and stored in a variety of software programming languages, may be publicly accessible or proprietary. A particular source code is typically built and tested on a particular computer hardware platform, and multiple teams may be involved in maintaining that one source code.

As the number of biological pathway models and our ability to accurately model any one biological pathway model increases, the challenge becomes how to integrate an ensemble of biological pathway models to build more *complex* models of cellular function. As an aside, this term complex needs to be discussed prior to proceeding. Any one biological pathway model within an integrated model may contain hundreds of species and a set of hundreds of resulting mathematical equations describing those interactions; however, this does not mean the model is necessarily complex.

For example, on a personal computer, super computer or even powerful handheld devices, hundreds of simultaneous differential equations can be solved; however, this does not mean that the model we solve is complex just because it has many equations. A complex system, on the other hand, may be complex even if the number of equations is small and apparently simple if the individual elements of the system have their own unique dynamic behavior. Such a system is said to be complex if it has multiple elements, which reveal different dynamic properties. This may occur, for example, when all system elements are continuous with concentrated parameters, but the model includes very fast and very slow parts [21]. Another example is a system where discrete parts interact with continuous sub-models of different speed and different kind such as an electronic circuit that contains integrated circuits as well as electro-mechanical parts such as relays and motors [22]. In other words, the model complexity has little to do with the model size.

Let us now consider a complex biological system: the interferon (IFN) response to virus infection. This integrated system involves various biological pathways, each of which is a unique domain of knowledge and effort for modeling. Figure 7 illustrates the four key biological pathways involved in this complex integrated system. Different research teams worldwide develop each pathway. At the lower left of this figure is the *virus infection pathway*. This biological pathway model simulates the virus infection of a cell and results initially in the up regulation of IFN-Beta, a critical signaling protein; and later on, results in the up regulation of IFN-Alpha, another signaling protein.

A second biological pathway model is *IFN receptor signaling*, as shown in the upper left. This biological pathway model represents the interactions of IFN proteins, either IFN-Alpha or IFN-Beta, landing on cell receptors to trigger the activation of other proteins within the cell's cytoplasm to up regulate IRF-7, an interferon regulatory factor. A third pathway is the IFN *amplification cycle*, as shown in the upper right. This biological pathway model simulates the production of increased amounts of both IFN-Alpha and IFN-Beta, which results from the by-product of virus infection with IRF-7. A fourth pathway is SOCS1 *regulation*, shown in the lower right. This pathway serves to regulate the production of IFNs by inhibiting the IFN receptor-signaling pathway.

The ensemble of all of the biological pathways depicted in Fig. 7, if integrated, can provide an integrative model of IFN response to virus infection. Each biological pathway is a contribution of different research teams across three continents of North America, Asia, Europe, and four countries: China, Russia, United States and Japan. Any individual biological pathway model within this ensemble does not have thousands of equations but the activity to integrate these four models to create one new model is unequivocally a complex problem for a number of reasons. First, based on the literature, not all of the biological pathway models depicted in Fig. 7 have source codes for their models. Second, the biological pathway models were built using different software programming languages. Third, each team developed their biological pathway models on different hardware platforms. Fourth, each of the biological pathway models exhibits different dynamic properties (e.g. different time scales). To integrate these four models is a complex problem. Such a problem is representative of most cellular functions that involve multiple biological pathways, which need to be connected to build a larger model.

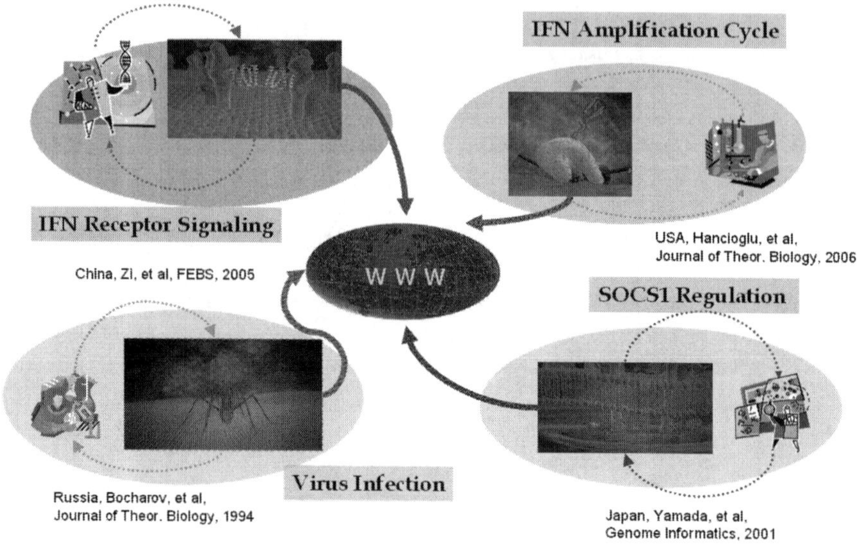

Fig. 7 Scenario of scientists performing systematic experiments to produce biological pathway models, which are published and made available over the WWW

2.2 Complexity of Maintaining an Integrated Model

The above discussion outlined the complexity of creating an integrative model of cellular function from combining various biological pathway models. This complexity is only one part of the problem. Another critical function is: the *maintenance* of the resulting integrated model. In the scenario described in Fig. 7, different teams worldwide develop each biological pathway model. Each team provides a particular domain of knowledge. As discussed in a previous section, the development of any one particular biological pathway model is an iterative process of systematic experimentation combined with quantitative modeling, both supporting each other. The reality is this: *Any one biological pathway model is constantly undergoing refinement.*

This means that the maintenance of a combined set of biological pathway models can be nearly impossible if there is no easy mechanism to receive and incorporate the updates from each biological pathway model; otherwise, the integrated model's accuracy is only good as its latest update. Since each model is developed in different mathematical and software representations, the integration and maintenance is made even more complicated.

2.3 Integration of Biological Pathway Models

Thus far, we have used the term, *integrated model* without formally defining it. Moving forward in our discussion, an integrated model refers to a group of biological pathway models executing together that have the ability to affect each other's computations. Based on the previous discussion, the question arises as to how does one effectively integrate an ensemble of biological pathway models and maintain them to ensure reliability.

One approach is to avoid the problem entirely and take a completely different approach: develop from scratch an entirely new biological pathway model that encompasses the multiple phenomena across each individual biological pathway model. In essence, create one large biological pathway model. The time and expense, however, involved in developing such a model is prohibitive. The design and implementation of a model requires a combination of software engineering skill and domain expertise. In addition, it involves an extensive amount of verification and validation, requiring the iterative process of systematic experimentation and quantitative modeling as previously discussed.

Another approach is to acquire and reuse the source codes from each biological pathway model, and merge them through some mechanism to create one large biological pathway model. While this process may be appearing easier than the previous one, it may not be so. The reuse of software is a key principle of software engineering and is usually achieved by developing a set of simple components, or modules that can be combined in different ways to create more complex components. Ideally scientists would connect their biological pathway models by integrating the existing source codes, treating them as modular pieces that can be easily and quickly plugged together. This, however, can be a very difficult task when the legacy source codes themselves may be poorly understood, and more than likely, were not originally designed to be integrated, and may be written in different software programming languages for different hardware computing platforms.

Despite the potential benefit of building new models from existing ones, integrating pathway models is not a common practice in systems biology community because of the difficulties inherent in working with source codes. Reusing source code in general is difficult for many reasons. Not only are programs difficult to comprehend (a necessary part of any software reuse) but the task of identifying useful source code fragments and integrating these source code artifacts that were not designed for reuse is challenging [22–24]. This is especially true for source codes written in unstructured languages and languages that make extensive use of global data e.g. Fortran [22]. Reusing source code is particularly difficult because biological pathway models are a unique class of computer programs whose design and use is intertwined with a great deal of domain-level theory outside the model code itself [25].

In short, the amount of time and expense required to understand the legacy source codes of each individual biological pathway model, prior to merging them may be more costly than starting from scratch. Moreover, once the integrated model is created, the next problem becomes the complexity of maintaining the resulting integrated model, since changes will no doubt take place in the originating biological pathway model's source codes, as they are refined and enhanced, through systematic experiments and modeling.

Perhaps a better way is to integrate biological pathway models in a decentralized manner such that the integrated model functions as one whole, while any of its component biological pathway models can continue to be owned and maintained by its original authors. If this approach is taken, effort will be required to build a new messaging architecture that enables disparately produced biological pathway models to interface, obviating the need to explicitly integrate the source codes. Such an infrastructure would allow scientists to quickly prototype those integrated models. This thesis is motivated to create such a computational infrastructure to integrate biological pathway models.

2.4 Research Question

The above sections should have clarified to the reader that biology is fundamentally an experimental science. The development and understanding of genes, proteins, and protein–protein interactions, or just any one element along the path to modeling the whole cell is difficult, time-consuming and complex, requiring the collaborative effort of multiple teams of scientists worldwide. The question then becomes: *How can we build larger models from smaller models in a scalable framework to support whole cell modeling given the reality of biology—an experimental science?*

3 Prior Work

Integrating biological pathway models is becoming an important area of research for advancing the field of systems biology. In the next section, we review the key factors that are driving this need and some recent efforts to create integrative models of biological systems. In the third section, we survey general computational architectures for integrating models. The fourth section specifically surveys the current approaches for integrating biological pathway models in the field of systems biology. We conclude this chapter by summarizing the current approaches in tabular form. We also provide a discussion on the critical weakness limiting the integration of biological pathway models.

3.1 Movement to Integrate Biological Pathway Models

There is a worldwide movement in the computational systems biology community to find powerful ways to integrate the growing number of biological pathway models. This movement is being driven by a transition from diagrammatic representation of pathways to quantitative and predictive mathematical models, which span time-scales, knowledge domains and spatial-scales [26–28]. This transition is being accelerated by high-throughput experimentation which isolates reactions and their corresponding rate constants [1]. Vast amounts of information is now available at the level of genes, proteins, cells, tissues and organs, which requires the development of mathematical models that can define the relationship between structure and function at all levels of biological organization [29].

Systems biology aims to provide a comprehensive quantitative analysis of the manner in which all the components of a biological system interact functionally over time [30]. Such an objective is pursued by an interdisciplinary team of investigators [31]. A significant computational challenge is how we can integrate such sub-cellular models running on different types of algorithms to construct higher order models [32].

Biological pathways, including metabolic pathways, protein interaction networks, signal transduction pathways, and gene regulatory networks, are currently represented in over 220 diverse databases. These data are crucial for the study of specific biological processes, including human diseases. Standard exchange formats for pathway information, such as BioPAX, CellML, SBML and PSI-MI, enable convenient collection of this data for biological research, but mechanisms for common storage and communication are required [33]. However, one the greatest challenges in establishing this systems approach are not biological but computational and organizational [34]. The critical need across all domains of molecular and cell biology is to effectively integrate large and disparate data sets [35].

Vigorous interest in understanding the dynamic aspects of cellular networks is also another driver in the development of integrative techniques for biological pathway models [36, 37]. Such explorations could provide insight into the mechanisms of healthy and diseased cells, as well as a better understanding of how system-level or whole-cell properties emerge from intracellular interactions of molecular components [38]. Moreover, understanding dynamics at the global network level seems to be now a reachable goal, which has motivated the growth of systems biology. Also, it is commonly admitted that the study of the network dynamics is able to enlighten the function of genes and groups of genes [39].

A central question now confronting virtually all fields of biology is whether scientists can deduce from this torrent of molecular data how systems and whole organisms work. All this information needs to be sifted, organized, compiled, and—most importantly— connected in a way that enables researchers to make predictions based on general principles [40]. Mapping protein interactions and transactions (such as phosphorylation, ubiquitination, and degradation) within a cell or organism is essential to developing a molecular understanding of physiology. Over the past decade, protein interaction mapping has evolved from low throughput manual screens to systematic interrogations of entire proteomes [41]. Reconstitution of biochemical and biophysical processes from 'minimal systems' of proteins has built confidence those top-down and bottom-up approaches to biology meet somewhere in the middle.

Systems biology has sought to integrate these results and data to reverse-engineer an understanding of biological network function and dynamics. The infrastructure for storing and disseminating information on biological systems, and for modeling them, has grown concurrently. In turn, this allows the rapid access and cross-comparison of information that is critical to establishing data quality and creating interoperability standards that will enable biologists to leverage their efforts and build scalable systems [42].

3.2 Existing Methods

Two critical elements need to be carefully assessed when selecting a modeling approach for any dynamic system: (1) the *level of abstraction* and (2) the *methodology of implementation*. In determining which level of abstraction and which methodology of implementation to use, the notions of *tractability*, *scalability* and *accuracy* are some of the important selection criteria, among others. Tractability is measured by the time and expense needed to design, implement, and test and assess the viability of the modeling approach. Scalability is determined by the ease with which the modeling approach can integrate new components at a particular level of abstraction. Accuracy is determined by the ability of the modeling approach to yield results, which match that observed in nature. Different users of the modeling approach will have these and other qualitative criteria in determining which modeling approach are the most optimal for their particular needs.

Currently there are two existing methods towards building whole cell models. The first method proposes to use first principles (ab initio) and large-scale computing, as has been applied in other fields such as climatology, particle dynamics, etc. to build a whole cell model. The second method involves downloading and accessing existing models and manually integrating their source codes together by hand to create one monolithic software program: *the monolithic approach*. A variation on this approach is to use semi-automation tools that help one to automatically read and integrate source codes together to create one monolithic software program.

3.3 First Principles—Ab Initio

There are various choices for which level of abstraction to use in modeling the cell. In Fig. 8 four potential abstractions are illustrated: quantum, atomic, biological pathways, and organelles. In the first principles approach, one could start at the atomic level of abstraction and solve the time-dependent Schroedinger equation (quantum mechanics, QM) to quantify the dynamics of the whole cell [43]. Such an approach would lead to a detailed understanding of the role that atomic level interactions play in determining the fundamental biochemistry of the whole

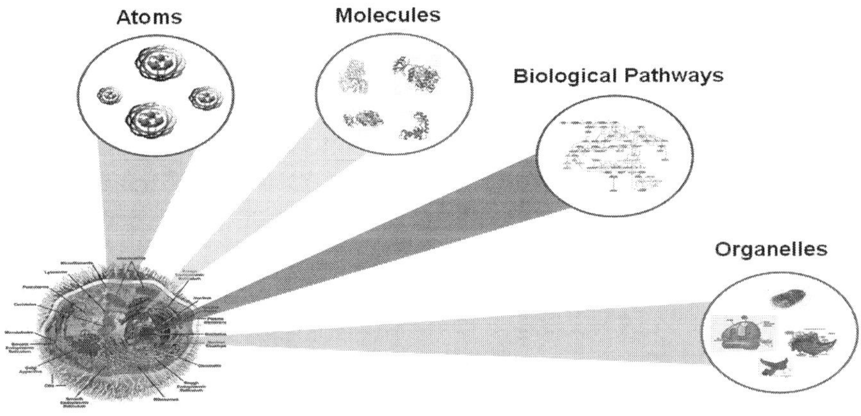

Fig. 8 Various levels of abstraction in modeling the whole cell

cell. The difficulty in using QM, for example, is that the vast range of length and time scales, from a nitrous oxide molecule to an organelle, makes the QM solution both impractical and useless [43].

It is impractical since there are too many degrees of freedom describing the motions of the electrons and atoms, whereas in the functioning of a cell it may only be the rate of transfer across some membrane. The complexity of QM limits its applications to systems with only 10–200 atoms (depending on the accuracy), leading to distance scales of less than 20 Angstroms and time scales of femtoseconds. Even the simplest protein in the cell contains over 1000 atoms. While the atomic level abstraction offers high level of accuracy, the level of abstraction is not scalable as the addition of each new atom increases the computational needs exponentially. This potential solution is also impossible, today, as the computing power needed to model the cell using this level of abstraction does not exist.

The first principle method therefore attempts to leap frog some of the steps in modeling the cell by using the laws of physics to model the cell versus experiments, which are the basis of biology. Another choice of abstraction in using the first principles method is at the molecular level, where Newton's equations are used rather than Schroedinger's to model molecular dynamics (or MD). Where in QM the solution is determined by averaging over the scale of electrons to describe the forces on atoms, in molecular dynamics (MD), one calculates an average over the dynamics of atoms to describe the motion of large molecules. While MD provides the ability to predict the dynamic interaction of molecules ab initio in an accurate manner, this level of abstraction is neither tractable nor scalable for modeling the whole cell since biological molecules such as proteins have far more atoms, degrees of freedom and numbers of states not encountered in other engineering fields where the species and interactions are well-defined [44]. We consider the simple problem of modeling the interaction of two proteins to demonstrate the intractability of using MD for whole cell modeling.

Consider the interaction of two proteins A and B to form the complex AB. We assume that each protein has 100 amino acid residues and each residue has three states (e.g. alpha, beta and other). In MD to model this simple interaction, two key calculations need to take place: 1. Thermodynamic, and 2. Kinetic in order to find the most likely transition state of protein A and protein B combining to produce complex AB [45]. For the thermodynamic calculations, MD requires the calculation of thermodynamic properties such as entropy which requires the need to evaluate all the possible states and associated probabilities of protein A, protein B and the complex AB. Protein A will have 3^{100} possible states (100 residues, each of which can be in three possible states), protein B will have 3^{100} possible states, and the complex AB can have up to 3^{200} (since AB is a combination of A and B) possible states. Just performing this calculation to determine the states and associated probabilities using modern computers is impossible and therefore intractable.

The kinetic calculation requires the identification of an appropriate reaction coordinate by computing the relative energies (or probabilities) for all the conformations along this reaction coordinate. In this case, it will require determining all the possible conformations of A and B that are at the energy of the activated complex, denoted by A' and B', (a higher energy than the energy of A and B); then determining all the possible conformations of A and B within complex AB stage, denoted by A'' and B'', (at an energy lower than that of A and B); and then finally determining all the conformations of AB complex. The kinetic calculation then attempts to link the most probable conformations starting with A and B, then A' and B', then A'' and B'', and finally the AB complex to calculate the reaction coordinate. These sets of multiple calculations using atom-by-atom MD to determine the reaction coordinate to solve even the simple molecular interaction of two proteins is intractable using modern day computers [45]. Moreover, this level of abstraction is not scalable as the number of interactions, number of proteins, and number of atoms per protein increases. In summary, while MD has powerful applications for determining protein conformations, it is not viable for whole cell modeling where hundreds of thousands of proteins are involved in millions of molecular interactions. Therefore, neither QM nor MD, using the laws of physics, offers tractable approaches to modeling the whole cell.

3.4 Biological Pathways as Modules

Another approach towards modeling the cell is to consider biological pathways as being the elemental modules from which complex cellular functions and the whole cell can be modeled. In this section, we present various viewpoints in the existing literature that supports such an approach. Biological systems are thought to have large number of parts almost all of which are related in complex ways [46]. Functionality emerges as the result of interactions between many proteins relating to each other in multiple cascades and in interaction with the cellular environment.

By computing these interactions, it can be used to determine the logic of healthy and diseased states [47]. One way to model the whole cell is through bottom up reconstruction. Such bottom up reconstruction, for example, of the human metabolic network was done primarily through a manual process of integrating databases and pathway models [48].

It is possible, for example, to regard signaling networks as systems that decode complex inputs in time, space and chemistry into combinatorial output patterns of signaling activity [49]. By treating biological pathways as modules our minds can still deal with the complexity. In this way, accurate experimentation and detailed modeling of network behavior in terms of molecular properties can reinforce each other [50]. The goal then becomes that of linking kinetic models on small parts to build larger models to form detailed kinetic models of larger chunks of metabolism, and ultimately of the entire living cell [20]. The value for integrating pathways is that it was found that the integrated network shows emergent properties that the individual pathways do not possess, like extended signal duration, activation of feedback loops, thresholds for biological effects, or a multitude of signal outputs [51]. In this sense, a cell can be seen as an adaptive autonomous agent or as a society of such agents, where each can exhibit a particular behavior depending on its cognitive capabilities.

Unique mathematical frameworks will be needed to obtain an integrated perspective on these complex systems, which operate over wide length and time scales. These may involve a two-level hierarchical approach wherein the overall signaling network is modeled in terms of effective "circuit" or "algorithm" modules, and then each module is correspondingly modeled with more detailed incorporation of its actual underlying biochemical/biophysical molecular interactions [52]. The mammalian cell may be considered as a central signaling network connected to various cellular machines that are responsible for phenotypic functions. Cellular machines such as transcriptional, translational, motility, and secretory machinery can be represented as sets of interacting components that form functional local networks [53].

As biology begins to move into the ''postgenomic'' era, a key emerging question is how to approach the understanding of how complex biological pathways function as dynamical systems. Prominent examples include multi-molecular protein ''machines,'' intracellular signal transduction cascades, and cell–cell communication mechanisms. As the proportion of identified components involved in any of these pathways continues to increase, in certain instances already asymptotically, the daunting challenge of developing useful models—mathematical as well as conceptual—for how they work is drawing interest [54].

Multi-scale modeling is essential to integrating knowledge of human physiology starting from genomics, molecular biology, and the environment through the levels of cells, tissues, and organs all the way to integrated systems behavior. The lowest levels concern biophysical and biochemical events. The higher levels of organization in tissues, organs, and organism are complex, representing the dynamically varying behavior of billions of cells interacting together [55]. Biological pathways can be seen to share structural principles with engineered networks, along with three of the most important shared principles, modularity, robustness to component tolerances, and use of recurring circuit elements. [56].

An important attribute of the complexity pyramid is the gradual transition from the particular (at the bottom level) to the universal (at the apex) [57, 58]. Others have recognized that one can build cellular-like structures from a bottom up approach [59]. Integrated models would represent the most compact, unambiguous and unified form of biological hypotheses, and as such they could be used to quantitatively explore interrelationships at both the molecular and cellular levels. [60]. At this time, for instance, the computational function of many of the signaling networks is poorly understood. However, it is clear that it is possible to construct a huge variety of control and computational circuits, both analog and digital from combinations of the cascade cycle [61].

3.5 Integrative Modeling Efforts

As discussed in the previous section, systems biology is determined to find new ways to integrate biological pathway models to build larger systems. Neuroscience, for example, seeks such integration of computational models for better understanding of different signaling pathways in neurons [62].

In the area of metabolism, researchers have created comprehensive mathematical descriptions of the cellular response of yeast to hyperosmotic shock. Their model integrates a biochemical reaction network comprising receptor stimulation, mitogen-activated protein kinase cascade dynamics, activation of gene expression and adaptation of cellular metabolism with a thermodynamic description of volume regulation and osmotic pressure [63].

The IUPS Physiome Project is an international collaborative open source project intended to provide a public domain framework for computational physiology, including the development of modeling standards, computational tools and web-accessible databases of models of structure and function at all spatial scales and across all organ systems [64].

For the first time, kinetic information from the literature was collected and used to construct integrative dynamical mathematical models of sphingolipid metabolism [65]. In another example, a model of 545 components (nodes) and 1259 interactions representing signaling pathways and cellular machines in the hippocampal CA1 neuron were combined. Using graph theory methods, this effort analyzed ligand-induced signal flow through the system. Specification of input and output nodes allowed them to identify functional modules. Networking resulted in the emergence of regulatory motifs, such as positive and negative feedback and feed-forward loops, that process information [53].

Oda et al. [66] have developed a complete map of the macrophage pathway. In this example, 234 published manuscripts were reviewed and 506 reactions were integrated within the single centralized software framework of Cell

Designer [67]. No models were integrated in this case; rather the work produced a large and complex monolithic diagram interconnecting the various biological pathways.

These examples demonstrate current efforts to integrate models to gain greater insight into a particular area of biology. The results seem promising, and such efforts are only growing. There is an also an equally growing need for foundational tools and architectures that support the continued development of such integrated models in a far more scalable manner. This has led to the development of many new tools such as Cell Designer, which aim to offer an easy-to-use interface for linking biological pathway models. In the next sections, we first survey general techniques that are used for integrating such models in order to gain a perspective for reviewing the more specific techniques in computational systems biology.

3.6 Generalized Architectures for Integrating Models

There are computationally two broad approaches to integrating multiple models: monolithic and messaging. The *monolithic approach* involves the creation of one monolithic source code resulting from the merger of the source code of the individual models. The *messaging approach* involves the need for no such merger, but creates a mechanism by which the necessary input and output data streams common across all models can be shared and transferred either statically or dynamically. Some have referred to this messaging approach as a *communicationsapproach* [22].

3.6.1 Monolithic Approach

There are three broad types of monolithic approaches for integrating models: *manual*, *semi-automated* and *module-based.*

Manual Monolithic Approach

The *manual monolithic* approach is process where the model integrator manually creates a single program or "file" by "cutting and pasting" the source codes or "wiring together" the pathway diagrams of individual models. Figure 9, illustrates the cutting and pasting of source codes from two biological pathway models Model A and Model B to produce Model C. A model integrator may alternatively wire together the pathway diagrams of two pathway models to produce a single pathway diagram using a visual design tool. Many find the manual monolithic approach easy to use. It offers full control to the model integrator of the source codes or pathway diagrams. (Fig. 10)

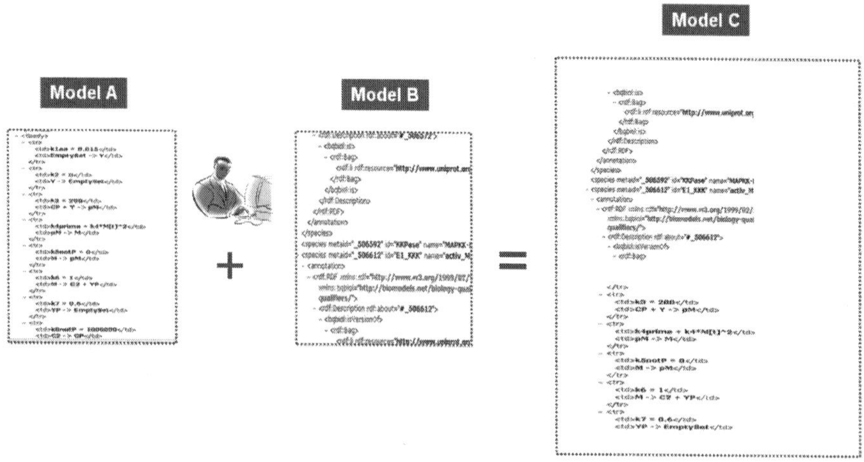

Fig. 9 Monolithic approach of cutting and pasting source codes of two models: Model A and Model B to produce a new source code of Model C

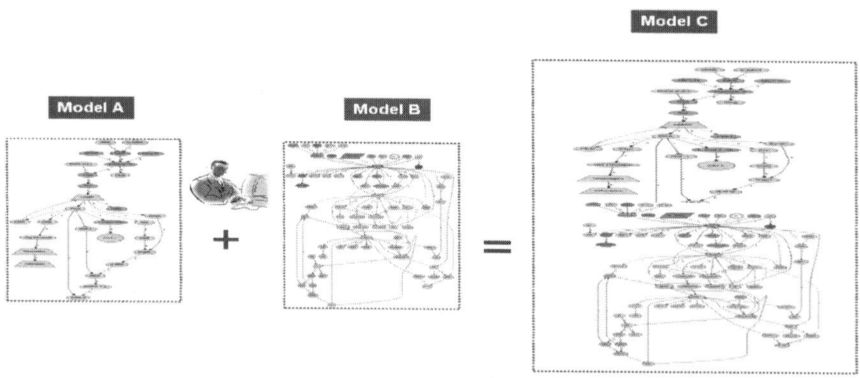

Fig. 10 Monolithic approach of wiring the pathway diagrams of two models: Model A and Model B to produce a new pathway diagram of Model C

The model integrator has control of all the coding details (control structure, memory allocation, data types, input/output file formats, etc.) [22]. Although this approach works, it has significant drawbacks. The individual performing the integration needs a complete and detailed understanding of the constituent models. In many cases, the source code is often difficult to obtain since legacy model codes are frequently complex, uncommented, and poorly documented [22]. The single integrated model's source code is also difficult to work with from a software engineering point of view (testing, debugging, verifying, updating, etc.) since it is much larger than its constituent model source codes, and improvements made to the original model source codes must be repeatedly made to each integrated model's source code as well.

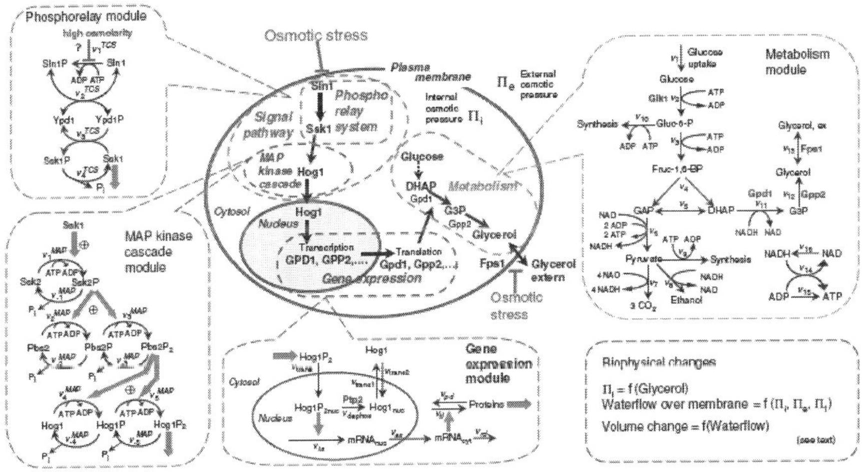

Fig. 11 An example of integrating models using the monolithic approach to integrate a model for the cellular function of osmoregulation [63]

Examples of recent published papers in using this monolithic approach are shown in Figs. 11 and 12. In Fig. 11 four models are integrated to produce on monolithic model for modeling the cellular function of osmoregulation [63]. In Fig. 12, three existing kinetic models are linked with one focusing on yeast glycolysis, a second extending this glycolytic pathway to the glycerol branch, and a third model introducing the glyoxalase pathway [20].

Semi-Automated Monolithic Approach

The *semi-automated* approach is a slight variation of the manual monolithic approach. In the *semi-automated* monolithic approach, software tools are used to accelerate the development of a single program from the individual model's source codes. In these approaches, a software tool, as illustrated in Fig. 13, combines together source codes or diagrams using some additional information to produce a single source code or diagram. Rarely do these tools produce the right source code, the first time. There is always manual intervention to review the initially produced output and then perform manual manipulations to produce the final output.

This semi-automated monolithic approach has the advantage of speeding up the initial merger process of source codes; however, the result varies based on the kind of semi-automation tool being used. In some cases, more effort is spent trying to get the tool itself working properly. Although this approach works, it too has the same significant drawbacks as the manual monolithic approach. One example of such a tool, SBMLMerge, is shown in Fig. 14 [68]. This tool merges two biological pathway models to produce one biological pathway model.

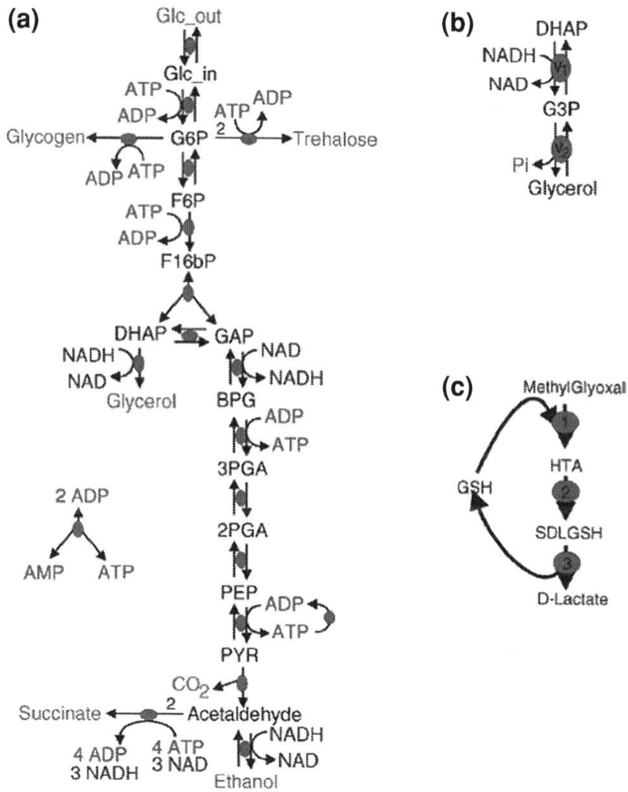

Fig. 12 An example integrating three pathways in a monolithic approach [20]

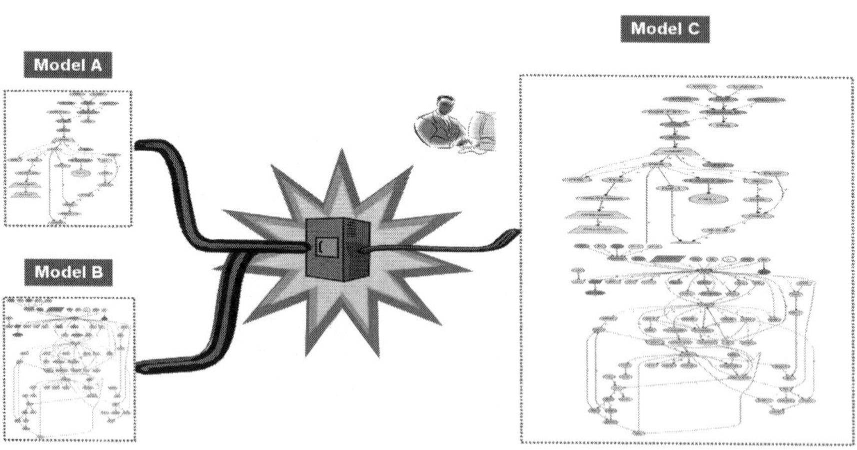

Fig. 13 Semi-automated monolithic approach of wiring the pathway diagrams of two models: Model A and Model B to produce a new pathway diagram of Model C

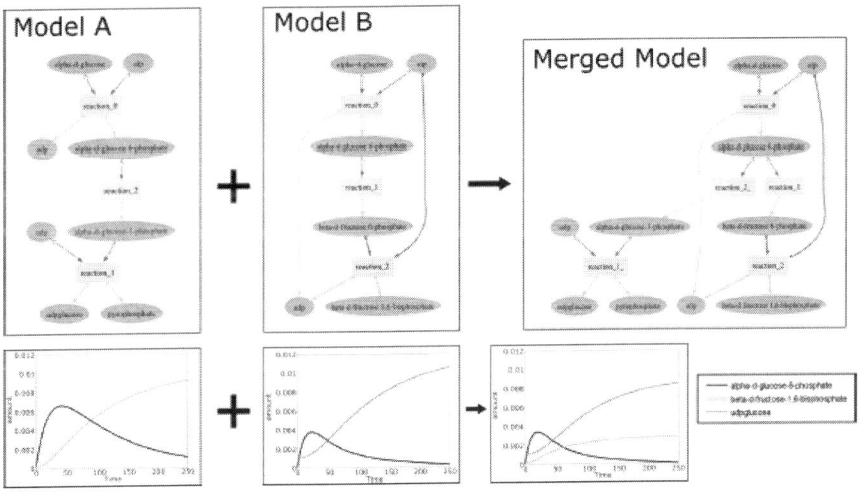

Fig. 14 SBMLMerge is an example of a tool that takes two biological pathway models and produces a merged model in a monolithic format [68]

For this tool to operate both biological pathway models need to be in SBML format and the tool produces one monolithic source code of the merged model in SBML format.

While these semi-automated tools may speed up the process, one disadvantage is that the model integrator does not have as much control of the resulting integrated source code, as they had in the manual approach. Some semi-automated tools insert new proprietary code and data structures along with a new format, which may take time to understand and debug. The model integrator, in that event, needs to not only have a complete and detailed understanding of the constituent models and their source code, as in the monolithic approach, but also now needs to understand how the tool for semi-automation itself works.

Module Based Monolithic Approach

The *module-based monolithic* approach addresses some of the limitations and drawbacks of the manual and semi-automated monolithic approaches. This approach offers the ability to employ reusable techniques for integrating models. The module-based monolithic approach results in the creation of single set of source code, but differs in that rather than decompose each model's source codes into blocks of source code designed for integration into another specific model's source code, the scientist decomposes each model's source code into software *modules*.

Modules are subroutines that are reusable and can be written with little knowledge of the other modules, and they can be replaced independently without significant changes to the rest of the program [22]. We use the term *reusable* to mean that a module can be used in a variety of different situations without any changes made to it. Each module possesses a standard interface for invoking and passing parameters. Connecting their interfaces then recomposes the modules. The interfaces for each module are generally simple and consist of a set of input data that must be supplied before the module can be executed, and a set of output data that is available upon completion [22]. The computation that a module performs is encapsulated and hidden within the module. These modules can be classified and organized into searchable libraries and the strict interfaces allow for automatic compatibility checking.

3.6.2 Messaging Approach

Due to the limitations of the monolithic approaches, scientists turned to approaches that obviate the need to produce either manually, semi-automated manually or even programmatically, one source code from the combined models and approaches that require a substantial programming. This led to the messaging approach. There are two types of messaging approaches: static and dynamic.

Static Messaging Approach

In the *static messaging* approach, the models remain independent programs and do not affect each other as they are executing. Any one model accepts as input a dataset, which may reside in any variety of formats, and executes to completion to generate an output dataset. That output dataset is then given to another model (perhaps after some transformation) which that model uses it as input and also executes through to completion. This process can then be continued with other models, and they can be executed concurrently if there are no dependencies between their datasets.

Architectures for supporting static messaging architectures offer tools that provide automated ways for the user to select models and datasets, and then specify the distribution of datasets [22]. For example, in one such architecture, called Le Select, a database-oriented approach is used in which both models and datasets are stored in geographically distributed databases and the user specifies the execution and data distribution through textual queries in a standard database query language [22]. Other architectures provide a visual interface to specify the order of execution of the models [69, 70]. These architectures typically have different user interfaces and different input/output data formats making them difficult to use, especially for non-specialists, requiring a significant investment of time to learn.

Some architectures provide a standard user interface to each model [71]. This requires that the original user interface source code be removed from each model and replaced with the common user interface source code. To address the issue of non-uniform data input and output formats, some architectures require the user to perform this data transformation manually between model runs while others require the user to change the model codes so that they use a standard data format [69, 70]. This requires that all the input and output source code be removed and replaced with source code to access the common database and use its data types.

Dynamic Messaging Approach

Using the dynamic messaging approach, the underlying model's source codes remain independently executing programs that interact only by exchanging data via message passing during execution. Architectures that use this approach can be classified by whether or not they include an independent application (a controller) that mediates the execution and messaging between the models.

The primary role of the controller is to transform exchanged data, which typically involves data type conversions, but they also sometimes control the startup of the models or track the global state of the integrated model as well. Architectures that do not include a controller are essentially libraries of data transfer and transformation routines customized for the data types and messaging styles needed by models. Architectures that do include a controller have messaging libraries that support direct model-to-model messaging as well as model-to controller messaging. In either case, these libraries often require the scientist to convert the model data into a standard data type, which is then communicated. All of these architectures require the scientist to perform an initial exercise of creating some mechanism of interfacing with each model through library calls in order to send or receive data.

The user then writes configuration files that specify which models are to execute, and the data that are to be sent and received. The messaging approach avoids the substantial model code rewriting required by the monolithic approaches, but the user still needs a complete and detailed understanding of the model codes in order to properly set up the correct interfacing for them [22].

In systems biology, for example, if one biological pathway model is written in MATLAB, and other model is in SBML, with each sharing four common variables that need to be communicated to solve the integrated problem, then the interface code is developed for the MATLAB model and for the SBML model. Once this interface code is written, both can transfer data. This interfacing process is often specific to a specific type of integration, so the interfacing process has to be repeated for each different type of interfacing.

4 Approaches to Integrating Biological Pathway Models

In this section, we will review current approaches to integrating biological pathway models within the field of systems biology. Two developments are predominant in addressing this integration problem: (1) the development of new software systems that allow the integration of multiple biological pathway models within a single centralized software framework, and (2) the development of common *standards* to define and code biological pathway models.

4.1 Software Systems for Integrating Biological Pathways

As discussed in the previous section on *Generalized Architectures for Integrating Models*, there are two broad classifications for approaches to performing such model integration: monolithic and messaging.

In systems biology, we propose another dimension of classification: *informational* and *computational*. Informational approaches are those that provide a way for integrating multiple sources of biological pathway information but do no computing with that integrated information. Computational approaches are those that are a superset of informational architectures by *also* provide the ability to perform integrated computations across the biological pathways. In the diagram below, we review the predominant software systems for integrating biological pathways and characterize them in this two-dimensional context of monolithic versus messaging and informational versus computational.

In the previous sections, we have surveyed a number of different architectures that can be used for integrating biological pathway models. In Table 1, we summarize these extant approaches based on our two-dimensional classification

Table 1 Summary of architectures for integrating biological pathways

Architecture	Monolithic/messaging	Informational/ Computational
Virtual CELL	Monolithic (manual)	Computational
Kinetikit/Genesis	Monolithic (manual)	Computational
Cell Designer	Monolithic (manual)	Computational
Jarnac/JDesigner	Monolithic (manual)	Computational
JSim	Monolithic (manual)	Computational
E-CELL	Monolithic (manual)	Computational
Gepasi	Monolithic (manual)	Computational
Jarnac	Monolithic (manual)	Computational
StochSim	Monolithic (manual)	Computational
PathSys	Messaging (static)	Informational
SBMLMerge	Monolithic (semi-automated)	Computational
CellAK	Messaging (module-based)	Computational
Cellulat	Messaging (module-based)	Computational
Cellware	Monolithic (module-based)	Computational
SigPath	Messaging (static)	Informational
Gaggle	Messaging (dynamic)	Informational
XPAUT	Monolithic (manual)	Computational
MATLAB	Monolithic (module-based)	Computational

methodology from the previous section. The 18 architectures summarized in Table 1 represent those architectures that support the integration of multiple biological pathway models. As noted earlier, some of these architectures support the integration of biological pathway model *information* but do not support the computing of the integrated models.

Most of the approaches are monolithic. Clearly within the monolithic approach, the module-based mechanism is an ideal way to construct new models if modules are available, but the approach is impractical for integrating existing models. Cellware and MATLAB offer a module-based monolithic approach; however, they are very difficult to perform the actual programming for this kind of application for a non-specialist, have a significant learning curve, and cannot produce code that is scalable. What is interesting to note; however, is that the most widely use architectures in the systems biology community is the monolithic approach.

The messaging approach allows existing models to be integrated with minimal changes to the model's source codes. Since we are interested in model reuse and scalability, we will focus on the messaging approach in this work. Moreover, we will ignore those architecture that are only *informational* for obvious reasons, since the do not support computation.

Based on Table 1, there are only two messaging-style architectures in the current systems biology community: CellAK and Cellulat. These two architectures offer an approach that lets biological pathway models be integrated without the need to explicitly integrate the source codes as in the monolithic approach. Neither of these approaches, however, performs dynamic messaging among the integrated models, which would be even more advantageous.

4.2 Weakness of the Current Approaches

As can be seen from the previous discussion, the predominant method for performing such integration is a *monolithic approach*, which involves creating one large biological pathway model through either a manual, semi-automated or module-based method that executes on a single computer.

There are many weaknesses, which make this approach not scalable. First, scaling to, for example, approximately 1,000 pathways—the level required to describe a single cell—would require a massive effort beyond the research and development expended to obtain the original individual pathways [72]. Second, each pathway represents a knowledge domain, and it would be essentially impossible to have one person sufficiently knowledgeable in all the scientific areas to understand each of these domains well enough to manually construct a single monolithic program.

Third, the monolithic approach does not provide a means for pathways from proprietary models to be used with other models that are open source. The monolithic approach wrongly assumes that the owners of any one model will be freely willing to share their models directly and will not protect proprietary information. The reality is that some models may be public and others, say ones owned by a pharmaceutical company may be private.

Fourth, most monolithic approaches support only one standard format. This means all other models need to be converted to that standard format. Fifth, and related to the previous weaknesses, the monolithic approach wrongly assumes that all models within an ensemble to be integrated are in the same format. The reality is that, while standards efforts are underway, models are coded in software platforms convenient to the author; and more practically any one platform is not capable of modeling all biological pathways. In the [66] example, all pathways had to be constructed in SBML and the integration had to be performed within the centralized framework of Cell Designer. Thus, this monolithic approach demands that in order to model the whole cell, all pathways would have to be placed into SBML or one common standard and then integrated within the framework of a centralized software system such as Cell Designer. Given the reality of standards adoption as aforementioned and the existence of 136 different software systems such as Cell Designer, a monolithic approach does not provide a scalable method to integrate multiple biological pathways to model the whole cell.

Sixth, the monolithic approach also wrongly assumes that all the models will run on the same computer and/or the same hardware platform. Different models may run on only certain hardware platforms, and more than likely were optimized and tested to run on particular hardware systems. This will become more important as the size and complexity increase, and special coding and computational accelerators are used to control the computational time.

Seventh, the monolithic approach assumes that all models reside in the same geographical location and ignores that biologists, even in a particular domain area of research such as immune response or cell motility, are distributed across laboratories worldwide. While they may build models within the same software platform and on the same hardware environment, the models themselves may be resident at a completely different location.

Eighth, the monolithic approach offers no real viable or practical mechanism to *maintain* the single large body of software code emerging from the merger of the software codes of the individual models. Consider four individual models that have been merged using the monolithic approach, and consider what process the author of the monolithic model will need to employ to track and maintain updates and changes to any of the four individual models. To any experienced software development manager, this is known as a *change management nightmare*. The reality is that any model may change constantly due to any one of several reasons including new advances in measurement, corrections to rate constants or identification of new species in a particular pathway.

Ninth, the monolithic approach, since it is centrally managed and maintained, places a new burden on the authors of the integrated model: that they become experts in the multiple domains represented by the individual pathway domains that were merged. This is nearly impossible.

Tenth and finally, many of the monolithic approaches attempt to enforce "standards" as a way to further reinforce a "monotheism". This centralized approach that would not fare well compared with more democratic, community-based approaches that understand and include research-driven development efforts. Creating a rigid standard before a field has matured can result in a failed and unused standard, in the best of circumstances, and, in the worst, can have the effect of stifling innovation [73].

4.3 What is Necessary?

What is needed is a method that directly addresses this integration and scalability problem by providing a parallel and distributed architecture allowing any individual pathway model to exist in any format and across different computers. Such a method would obviate the need to manually load, understand and interconnect each individual pathway, as is required in monolithic systems. It is clear from the above literature review, the common approach today is a monolithic approach in which computational biologists seeking to create an integrate model "cut and paste" source codes to create on monolithic model.

More specifically, of the two recent papers that provide examples of integrating biological pathway models, both involve the merging of SBML models using a manual monolithic approach [20, 63]. Thus, from a short-term perspective, even an architecture, which allows the integration of distributed SBML models, would be of huge benefit. SBMLMerge offers a tool that is a semi-automated monolithic approach, while valuable for those seeking to "cut and paste" models faster, it is not a scalable solution, since the resulting code base still cannot be maintained as the models change overtime, requiring re-integration each time every time the elemental models change.

A more advantageous solution would be an architecture that could, with minimal effort and no programming, *connect* or *couple* the codes of multiple SMBL models, linking their computations. Such a tool would be of immense benefit. If this same architecture could allow for the integration of models also not written in SBML, but in any other format, it will accelerate the development of complex models, versus waiting for all extant models to be converted and curated in SBML. This too will be particularly useful, given the reality that most of the encoded models available in scientific publications or on the Internet are not in a standard format. Of those that are encoded in a standard format, it turns out that most actually fail compliance tests developed for these standards [74]. In fact markup languages for model encoding such as SBML may not be always the only way for modeling biological pathway models, there are others that provide a range of descriptive and analytical powers. As the field matures, there will be a wider uptake of these alternative approaches for several reasons, including the need to take into account the great complexity of cellular organization [75].

From a software engineering perspective, each biological pathway model represents an individual software program, each with different inputs and outputs, written in different programming languages, by different developers, potentially distributed worldwide. Modeling the whole cell, therefore, can be likened to a large-scale systems integration problem.

The research goal of this effort, therefore, is to develop a new approach to integrate biological pathway models that overcomes the intractability and lack of scalability of existing approaches. In his seminal work, [76] has demonstrated that the amount of effort to develop such a large-scale system increases exponentially with the amount of additional personnel communications required to coordinate development personnel. The thesis will explore a method that resolves this bottleneck in personnel communications by allowing individual teams to own and manage their own pathway models without being involved in a massive project management effort to coordinate the integration.

5 The Platform

This section presents the platform and scalable architecture for integrating biological pathway models. This architecture is called *CytoSolve*.

5.1 Key Requirements

Ten key requirements are identified below to address the weaknesses of previous approaches.

5.1.1 Scalability

If the goal is to model complex cellular systems and eventually the whole cell, the architecture must be able to integrate new pathway models with the same ease as it is to integrate the first one. Scalability is measured by the ease in which additional models can be integrated. Recall that complexity of integration, from our earlier discussion, has little to do with the number of equations in any one model. Two models with numerous equations can be relatively easy to integrate if they are written in the same program, same time scales, in the same domain and developed on the same hardware platform.

5.1.2 Opacity of Multiple Knowledge Domains

The architecture must express the property of *opacity*. Opacity means that when one individual is integrating a particular pathway model into a pre-existing ensemble of integrated models, one should not have to know the details and inner workings of all other pathway models. Each pathway may represent a unique knowledge domain, and it would be essentially impossible to have one person sufficiently knowledgeable in all the scientific areas to understand each of the domains.

5.1.3 Support for both Public and Proprietary Models

The architecture must support both Public models (models where the source code is readable and available to all) and Proprietary models (models where the source code is inaccessible). The monolithic approach does not provide a means for pathways from proprietary models to be used with other models that are open source. Many pharmaceutical companies, for example, will not want to share the inner source code of their particular proprietary models; however, they are interested in coupling their models with other models to gain better understanding of a larger cellular process. Alternatively, researchers in the academic environment may wish to integrate their Public models with existing Proprietary models to learn some new aspect of science, but cannot currently due to confidentiality issues. By enabling a way to ensure protection of the source code of those Proprietary models, new industry and academic collaborations will be possible with far greater ease. Those with Proprietary models currently chose either not to follow standards to protect their models or if they do follow standards are unwilling to share their model code, which was the reason for the standard itself.

5.1.4 Support for Multiple "Standards"

The architecture must support any pathway model code in any format or "standard." While the architecture should support the integration of pathway models constructed in a standard such as SBML, for example, it should be able to communicate with models in any format.

5.1.5 Heterogeneity of Integration

At any time models may be in different formats. The architecture should support the ability to integrate, in real-time, models that are in different formats. Thus, if there are three models, even if one model is in SBML, another in MATLAB and a third in FORTRAN, the architecture should be able to integrate them with minimal to no effort.

5.1.6 Cross Platform Support

The architecture should allow models developed on different hardware and computing environments to be integrated with ease. Different models may run on only certain hardware platforms and more than likely were optimized and tested to run on a particular hardware system. It will be far easier to keep a model resident on a hardware platform for which it was designed and tested for versus having to recode or reconfigure it in any manner to a new hardware platform, which could prove to be very expensive and time-consuming.

5.1.7 Independence of Location

The architecture should support integration of models across geographical boundaries. While each model may be on different computers, they may also be physically at different locations anywhere in the world. The model should support protocols for communicating with models anywhere without regard to geographical location.

5.1.8 Ease of Maintenance

Any model integrated within the architecture should be able to dynamically change with little to no source re-coding efforts to incorporate changes for that model into the larger integrated model. In the monolithic approach, any change to an individual model typically requires significant recoding and retesting of the integrated model. In this new architecture, we want to avoid such a process. This requirement is extremely important to create a scalable architecture. The addition of new models should not require changes to any of the other existing models.

5.1.9 Decentralized Management and Distributed Control

Decentralized management and distributed control means that each model is maintained at the "local" level, not at a central level. The current monolithic approach requires centralized model curation, such as many of the existing model repositories. We want to support local management of models. Moreover, any creator of a model should be able to integrate their model from their local location to an ensemble of distributed models. This means that if one owner of a model has Model A and wishes to quickly test or integrate their model with a set of three other models: Model B, C and D. They should not have to download each of the other models to their local computer. With ease, the architecture should enable the owner of Model A to integrate with the other three models with little to no effort.

5.1.10 Hierarchical Support

Biological systems are systems of systems. This means that the architecture should support the ability for systems to be composed of other systems, while ensuring that each system satisfies the requirements herein.

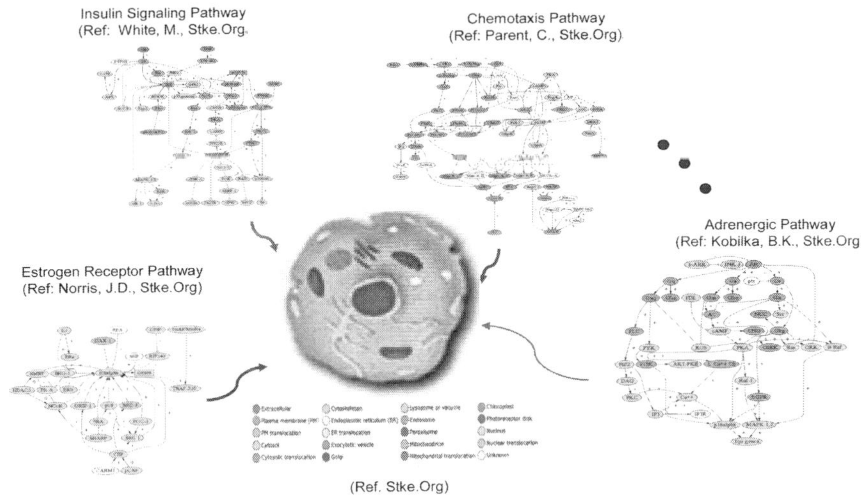

Fig. 15 A cell as an interconnection of biological pathways

5.2 Functional Specification

The above ten requirements will be used as the basis for defining the architectural design. We begin our functional specification process by abstracting the cell to be interconnection of biological pathways, as shown in Fig. 15. In this approach, we make the following key assumptions:

1. The cell or compartment is well-mixed. This means that a sufficiently long-time between reaction collisions takes place to ensure that each pair of molecules is equally likely to be the next to collide. This also means that the concentration of each species is high and transport essentially instantaneous.
2. The progress of the system only depends on previous state (e.g. Markov process).
3. Between cells and compartments, transport is slower and associated with an observable rate.
4. We treat each pathway model as a *black box*. This means the following:
   - Inputs and outputs are species concentrations
   - Changes in localization are represented by *compartments*
   - Species are defined by their compartments
   - Species can move through compartments
   - Species can inhabit one or more compartments
5. The cell can be modeled as an integration of biological pathway models. Recall, biological pathways are moving from diagrammatic representations, as shown in Fig. 2 above, to biological pathway models, and each biological pathway

model has internal parameters along with inputs and outputs which are the molecular species concentrations for the nth and $(n + 1)$th time step, respectively.
6. The cell can be modeled as an integration of biological pathway models. Recall, biological pathways are moving from diagrammatic representations, as shown in Fig. 2 above, to biological pathway models, and each biological pathway model has internal parameters along with inputs and outputs which are the molecular species concentrations for the nth and $(n + 1)$th time step, respectively.

Thus, modeling the cell therefore can be seen as an interconnection of biological of pathway models as shown in Fig. 15.

5.3 Platform Design

Based on the earlier discussions, it is clear that the monolithic approach is widely used primarily because there are no other real alternatives in systems biology today. The main problems with this approach have already been discussed. The messaging approach, which has not been fully developed for systems biology, except in the two cases of Cellware and Cellulata, is in the right direction. However, both of these methods use a static messaging approach. There are no known solutions, to our knowledge, which use a dynamic messaging approach that supports scalability as well as ease of maintenance. Our first decision is to pursue a dynamic messaging approach as the basis of the architecture. This dynamic messaging approach provides nearly all of the advantages to address the weaknesses of the monolithic approach and the static messaging approach, for the reasons previously stated.

5.3.1 CytoSolve: A Dynamic Messaging Approach

Based on the above discussion, we now introduce CytoSolve, a scalable architecture for integrating an ensemble of distributed biological pathway models. In Fig. 16 the layout of the architecture for supporting a dynamic messaging approach is illustrated. The layout of this architecture will meet all of the requirements of the specifications outlined in the previous sections.

The elements of this architecture are:

- Biological Pathway Models—These are the boxes with the input and output arrows along the outer edges of the diagram. Each biological pathway model is a computer software program.
- Internet—This is represented by the internet "clouds". This serves to show that the biological pathway models can reside anywhere geographically on the planet

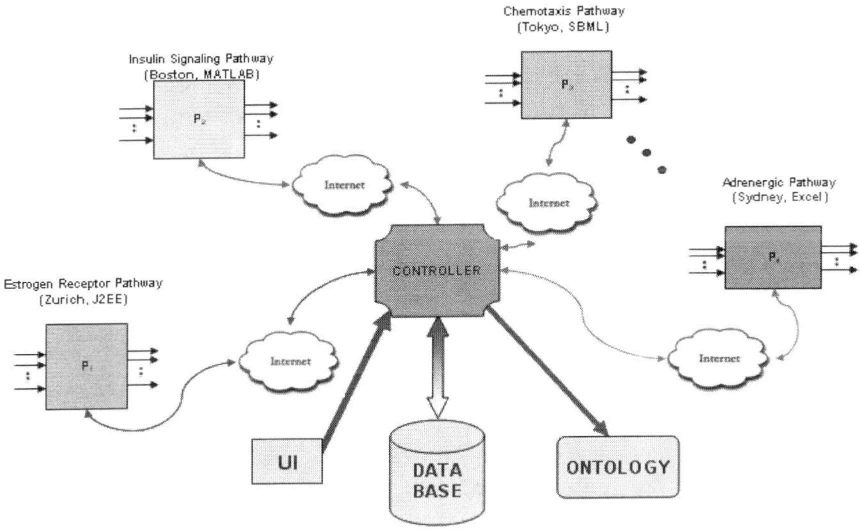

Fig. 16 CytoSolve—a dynamic messaging approach

and can use the internet for communication through the controller. This does not mean that we *have* to use the Internet. All models can be centralized on one computer and CytoSolve will communicate locally to each model.

- Controller—This module serves to coordinate the computational activities across the various models.
- Controller-to-Pathway Interface—These are the arrows in the diagram from the Controller to the Pathway, and represent the mechanism by which the controller communicates with each individual pathway model.
- User Interface—The user interface allows the user to specify which models will be included within the architecture.
- Ontology—This is the data which specifies characteristics of each model's input–output behavior to allow the Controller to effectively communicate across all models.

The key features of this architecture include an infrastructure that provides simple communications interface to each model, is distributed, Web-enabled, and automatically aggregates the models to build the integrated model. The architecture supports both distributed and parallel processing, while using a hybrid of shared memory and message passing. The shared memory is for tracking the species concentrations across all models in time. Message passing is used for remote communications between the controller and individual models. The architecture is built in an open environment supporting: (1) publicly available tools, and (2) emerging standards. The discussion of the details of the implementation, which is evolving, and currently in a web-based environment at www.cystosolve.com, is beyond he scope of this discussion.

5.3.2 CytoSolve: Solution Methodology

CytoSolve dynamically integrates the computations of each model to derive the species concentration of the integrated model. Details of the solution methodology are detailed in another publication [77].

In summary, the method works with the Controller performing initialization of the system by allocating memory storage for the computed species concentrations of each model and the integrated model. A component called the *Monitormonitors* the progress of each model's computation, during its initialization, accesses the initial conditions. Control is then passed to the a component called the *Communication Manager* which awakens all the models to start up and become ready to process a time step of calculation, and then invokes the Monitor.

The Monitor proceeds to invoke all models in parallel to execute a time step of calculation using the species concentration values of the integrated model at time step n, as the input to all models. Each model executes and computes one time step of calculation on its own Remote Server. Monitor tracks the progress of each model's completion. Once a model completes its computation, the output is stored, and the model then goes to *sleep* to optimize use of the Remote Server's CPU usage. By sleep, we mean that the model goes dormant until invoked again by the Communications Manager to process another time step of calculation. Once all models have completed their processing for a time step, the Monitor passes control back to the Communications Manager, and the Monitor itself goes to sleep.

Once all models have completed a time step of calculation, the Communication Manager invokes the *Mass Balance* component to dynamically couple the computations at time step n of each model to evaluate the integrated model.

6 EGFR Model Validation

CytoSolve is validated by comparing the solution it produces with the one generated by Cell Designer, a popular tool for building molecular pathway models in a monolithic manner. As a control, the Epidermal Growth Factor Receptor (EGFR) model published by Kholodenko et al. is selected for this comparison. Snoep et al. have authored the model into the SBML language. The *entire* EGFR model is loaded into Cell Designer and executed on a single computer. The same entire EGFR model, to test CytoSolve, is *split* into four models and distributed on four different computers, as shown in Fig. 17. CytoSolve and Cell Designer are run for a total of 10 s in simulation time.

6.1 Results

CytoSolve and Cell Designer produce near exact results as shown for two example species in Fig. 18. The results demonstrate the viability of CytoSolve's

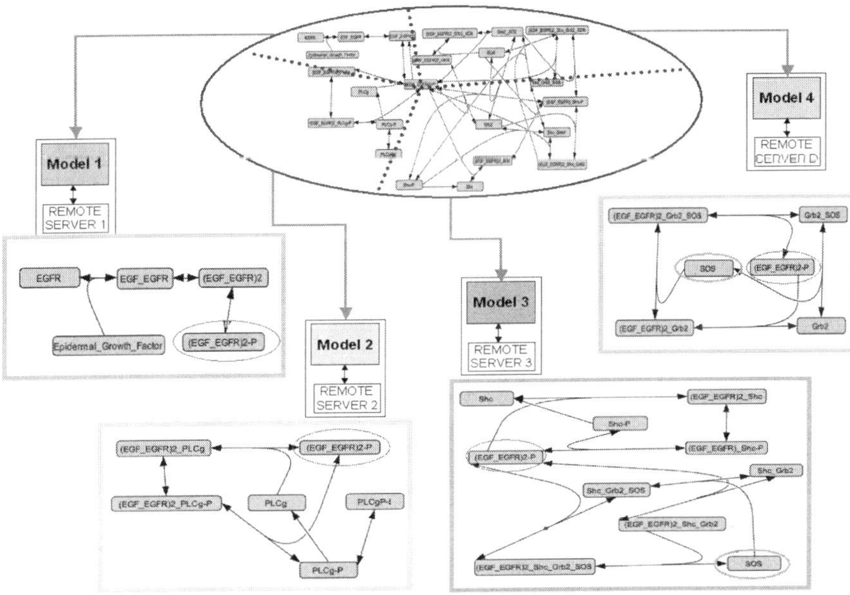

Fig. 17 Complete EGFR model of Kholodenko split on four Remote Servers for CytoSolve solving

unique distributed approach not only to solve problems that monolithic approaches are capable of solving but also to provide greater flexibility and scalability in integrating multiple biological pathway models, which monolithic approaches are incapable of doing. In CytoSolve, any *one* pathway can exist in any format on any computer, and there is no need to manually load, understand and interconnect each individual pathway, as is required in monolithic systems.

CytoSolve generated exact results to Cell Designer; more importantly, the integration of the four models in CytoSolve did not require any manual "wiring" as is needed by Cell Designer. CytoSolve's compute time was greater than Cell Designer; however, most of this compute time was due to Transmission Time. Since CytoSolve works in a distributed parallel fashion, its compute time is a direct function of the compute time of the largest pathway plus the associated Transmission Time and overhead for Controller Time to integrate. For Cell Designer, the compute time will be the compute time of the whole integrated pathway.

Finally, and perhaps equally important, is that managing a monolithic model, composed of other models, is a change management nightmare. Consider a small example of a monolithic model "cut and pasted" or concatenated from the four models of EGFR, aforementioned, and each model being published and created by different authors. Now, suppose once the monolithic model has been constructed,

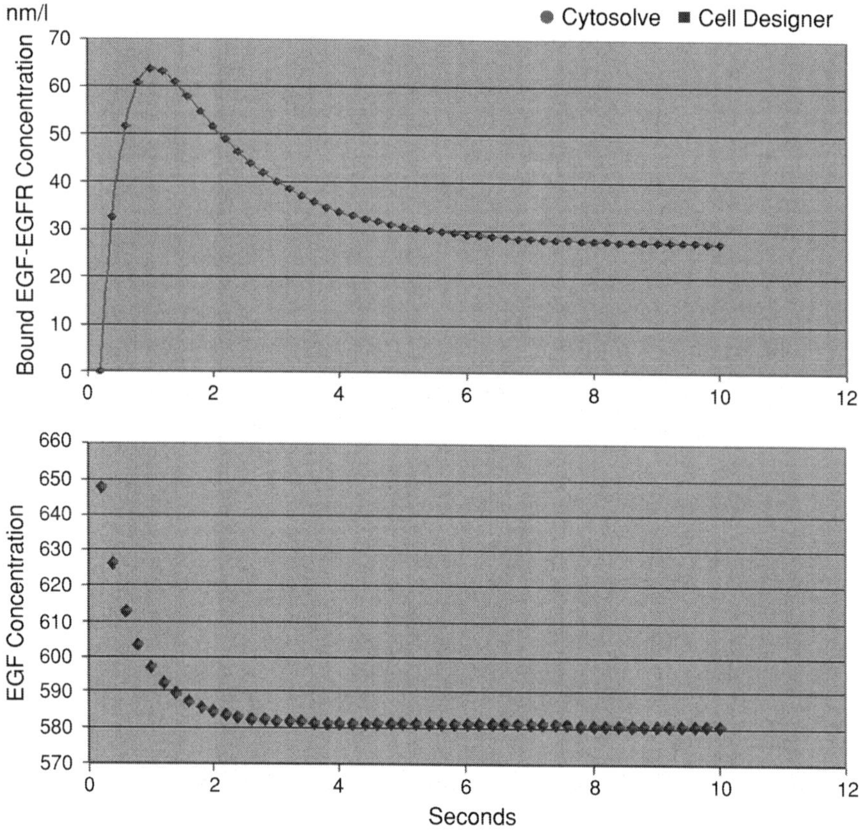

Fig. 18 Comparison of results from CytoSolve and cell designer for two species. **a** compares values of the EGF-EGFR species. **b** compares values of EGF concentration

that many months later, the authors of each of these models changes rate constants, pathway connections, etc., at that point the author of the monolithic model would have to rebuild the entire monolithic model, by instantiating changes from each author's model, which may be tenable for four models (possibly based on the complexity and domain specificity of each model). Modeling the whole cell while managing such changes across a suite of hundreds of such models will be untenable.

7 Solving an Unknown Problem Using CytoSolve

The immune system has many different types of cells acting together to protect the body against viruses, bacteria, and other "foreign invaders." Part of this protection includes the production of interferon (IFN), a protein that plays a special role in triggering the body's response. The following describes what interferon is and why it is so important to the immune system.

7.1 IFN Response to Viral Infection

The immune system consists of a complex network of cells, tissues, and organs all working in tandem to ward off infection and keep us healthy. This includes interferon, one of the proteins called cytokines, which are diverse and potent chemical messengers that can trigger the immune system to attack invading pathogens. Interferon signals neighboring cells into action and also interferes with how foreign cells grow and multiply.

In humans, IFNs also play a roles in cell growth, differentiation and immunomodulation. IFNs are divided into two groups depending on their molecular basis; type I IFNs (IFN-alpha and IFN-beta) are produced by a variety of cells following virus infection, and type II IFN (IFN-gamma) is produced by activated T cells and natural killer (NK) cells [78]. There are three classes of Interferon, alpha, beta and gamma. Interferon alpha and beta are produced by many cell types, including the infection-fighting T-cells and B-cells in the blood, and are an important component of the anti-viral response.

7.2 Elements of the IFN Response

The IFN response mechanism of the cell to virus infection is a core cellular function. There are four key biological pathways, which are involved to elicit IFN response to virus infection as shown earlier in Fig. 7:

- Up regulation of IFN-Beta
- IFN receptor signaling to produce IRF-7
- Virus amplification cycle to produce more IFN-Beta and IFN-Alpha
- Regulation and balancing by SOCS-1

7.2.1 Virus Infection Model

The virus infection pathway creates IFN-Beta as an initial response to virus infection. Scientists in Moscow, Russia in 1994 modeled this pathway [79]. The original code was written in MATLAB. In this pathway, the viruses injects single-stranded RNA into the host cell, which leads to the formation of double-stranded RNA. Double-stranded RNA triggers the activation of virus-activated kinase (VAK), which phosphorylates IRF-3. Phosphorylated IRF-3 is a transcription factor for the IFN-Beta gene. The expression of this gene results in the initial production of IFN-Beta.

7.2.2 IFN Receptor Signaling Model

The IFN receptor signaling produces IRF-7 as a preparation mechanism for the infected cell and neighboring cells. Scientists in China in 2005 defined this pathway model in *FEBS* [80]. Only a pathway diagram along with parameters exists for this pathway model, but no software code. IFN-Beta (or IFN-Alpha) lands on the IFNAR receptor to initiate the up regulation of IRF-7 which is a critical protein for signaling the cell itself as well as neighboring cell of the virus infection. The binding of IFN with the receptor leads to the STAT protein being phosphorylated. The phosphorylated STAT forms a homo-dimer and becomes the transcription factor for IRF-7 gene after binding to IRF-9, which leads to expression of IRF-7. This signaling mechanism *prepares* the cell for further defenses by producing IRF-7.

7.2.3 IFN Amplification Cycle Model

The IFN amplification cycle is a critical step in the response to protect the cell from virus infection. A team of scientists from the America created a dynamic model of this pathway. The article was published in the *Journal of Theoretical Biology* [81]. They programmed the pathway in XPAUT which can be saved in SBML. The virus interaction with IRF-7 not only serves to up-regulate IFN-beta but also serves to up-regulate IFN-alpha.

7.2.4 SOCS1 Regulation Model

This biological pathway produces SOCS1 to regulate and balance the production of IFNs. Without this pathway, the additional levels of IFNs, beyond what is necessary to stop the virus infection, can itself have detrimental effects on the cell. Scientists in Japan in 2001 defined this pathway model in *Genome Informatics* [82]. They programmed the pathway in MATLAB. Here, JAK binds to the IFN receptor and forms the JAK-IFNR receptor complex. Once IFN binds to the receptor, the resulting complex associates with each other and forms a homo-dimer. This dimer undergoes phosphorylation, leading to a form as IFNRJ2*, which catalyzes the phosphorylation of STAT1. The phosphorylated STAT1 also forms a homo-dimer and acts as a transcription factor of SOCS1 gene. The resulting protein, SOCS1, inhibits the kinase activity of IFNRJ2 and is the key component of the negative feedback loop.

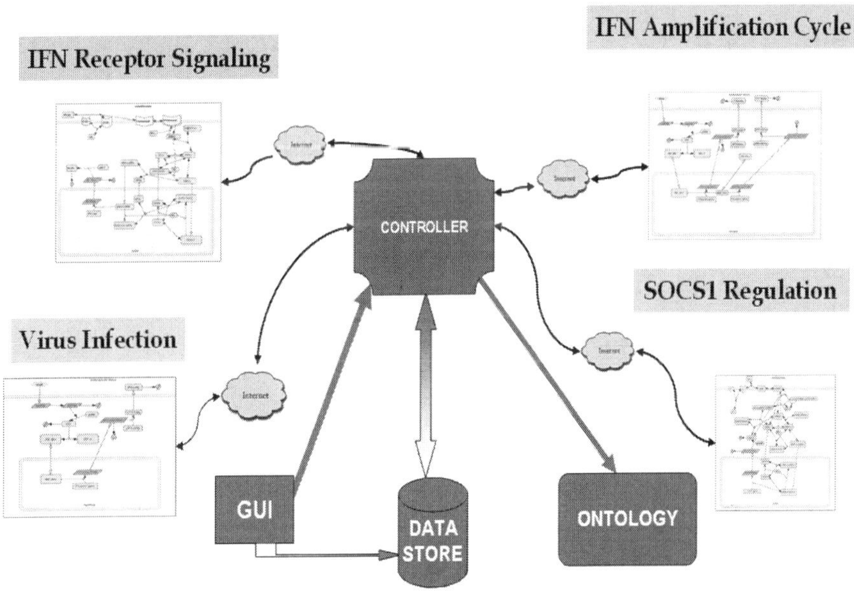

Fig. 19 CytoSolve approach to integrating the four models

In Fig. 19 below, each of the above models, as depicted is loaded into CytoSolve for dynamic integration.

7.3 Results

The summary results from the integrated model of IFN are shown in Fig. 20. Key molecular species presented in this figure are IRF-3, IRF-7, IFN-Beta and IFN-Alpha. This integrated model combines the four pieces of interferon pathway: virus infection leads to up regulation of IFN-Beta; IFN-Beta then results in the creation of IRF-7; the existence of IRF-7 then results in positive feedback to which increases a massive production of IFN-Alpha and IFN-Beta; finally, control of JAK/STAT signal transduction pathway by SOCS1 results in regulating and balancing the production of IFN-Alpha and IFN-Beta.

Close review reveals several important elements of the integrated model. First, during the first ~13 h (~50,000 s), the concentration of IRF-7 through time is a sigmoidal curve which reaches the steady state value of 0.7 nM. Second, during this same first ~13 h period, the concentration of IFN-Beta and IFN-Alpha slowly increases. What is interesting to note is that within the first 3.3 h (~12,000 s), the initial production of IFN-Beta is then followed by the production of IFN-Alpha. IFN-Beta is produced within the first 30–40 min (~2000 s to ~2500 s).

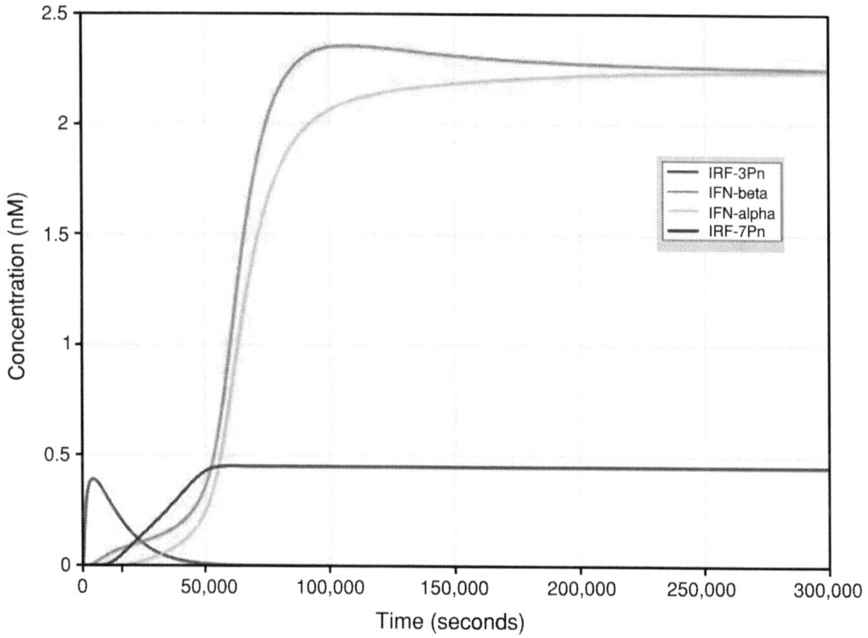

Fig. 20 Integrated model solution

The initial production of IFN-Beta, after the 40 min period and before the 3.3 h period is defined by a marked increase in IFN-Beta production.

Third, after the first ~13 h (~50,000 s) to ~25 h (~90,000 s), IFN-Beta and IFN-Alpha exponentially increases. Fourth, after ~25 h (~90,000 s), IFN-Beta and IFN-Alpha concentrations reach their maximum and gradually approach steady state due to the balance between positive feedback system and negative feedback control from SOCS1 activation.

The results of the CytoSolve solutions were tested with known experimental data. This revealed that IFN Beta production in the first 30–40 min, as predicted by the model, matches experimental data. In addition, IFN-Alpha begins production after a ~3 h delay time, and this verifies with the experimental approximated expected time required for the positive feedback cycle to start. Finally, both IFNs reach their peak in the ~20 h range as predicted by various experimental research [83–85]. Greater details of this integrative model of IFN are described in a forthcoming paper.

8 Discussion and Conclusions

This paper has introduced CytoSolve, a new computational environment for integrating biomolecular pathway models and a collaborative platform for computational molecular systems biology. The initial results from the EGFR example has demonstrated that CytoSolve can serve as an alternative to the monolithic approaches, such as Cell Designer. The solution of the IFN integrated model demonstrates CytoSolve's ability to solve new problems by merging multiple molecular pathway models.

The purpose of CytoSolve is to offer a platform for building large-scale models by integrating smaller models. Clearly, if we want to model the whole cell from hundreds of sub-models, each of which is owned by various authors (each making changes to their models), the monolithic approach is not scalable. CytoSolve's dynamic messaging approach offers a scalable alternative since the environment is opaque (treats each model as a black box) support for both public and proprietary models is extensible to support heterogeneous source code formats, and finally supports localized integration, a user can initiate integration from their own local environment.

CytoSolve is now available at http://www.cytosolve.com as on-line web computational resource. The portal offers researchers an environment to collaborate and integrate quantitative molecular pathways without needing to be a specialist on the computational architecture and neither an expert of all the multiple sub-models to be merged. Future work we will include a more sophisticated Ontology to manage nomenclature and species identification across all individual biological pathway models to be integrated by means of the web application, and automated searching for related biological pathways.

8.1 Future Work

This research may also serve as a foundation for several new areas of investigation including:

8.1.1 Spatial Scale Variation

At the present time, CytoSolve supports only computational models that represent one single pool of material or several distinct pools connected with specific transport relations. We have not considered changes in concentrations on a continuous spatial scale. We believe that the architecture, based on its modular approach and support for multiple compartments, can support varying spatial scales. However, more testing will have to be performed to understand the computation times required to fully support such spatial variations. The description language FieldML is available to support this process.

8.1.2 Adaptive Time Stepping of the Controller

All models are currently computed using a single adaptive time step, which is taken to be the fastest time step among the ensemble of models. This is not optimal, as some component models may be varying more slow than others. Additional effort is required to implement intelligent adaptive time stepping at the Controller level to observe the time scales of different models and invoke them only when necessary. Such an effort will result in improved computation time performance.

8.1.3 Implementation and Integration with Emerging Ontologies

CytoSolve has support for integrating other ontologies such as MIRIAM; however, future research needs to be done to fully integrate MIRIAM and other such ontologies. This effort will enable CytoSolve to support many more model formats with greater ease, leveraging standards that the systems biology community globally accepts.

9 Epilogue

In the Summer of 1978, I was 14 years old and had just completed my sophomore year at Livingston High School in New Jersey. That summer I had been fortunate to have been accepted into the Courant Institute of Mathematical Sciences gifted students program in computer science at New York University. During that Summer, I and 40 other of my fellow students learned eight different computer programming languages including FORTRAN, COBOL and PL/1 to name a few. The course that summer at NYU, and my deep interest in mathematics, in retrospect were important elements that supported my future work.

Swamy Laxminarayana, or Swamy as I called him, however, was my gateway into a completely new world from which I can trace nearly all of my accomplishments. Swamy exposed me to the world of pattern analysis and recognition at an early age. He was a friend above all; although I was nearly 30 years younger than him, he treated me as a colleague with great respect, genuine warmth, and love. My mom, Meena Ayyadurai, worked at the University of Medicine and Dentistry of New Jersey (UMDNJ) as a Systems Analyst while Swamy worked upstairs with Dr. Leslie P. Michelson in the Laboratory for Computer Science. My mom had first introduced me to Dr. Michelson with whom I had begun to build an electronic mail system. In fact, that electronic mail system was the worlds first E-Mail System, for which I received recognition by the Westinghouse Science Award in 1981.

I recall while I was in the midst of building that E-Mail System in 1980, Dr. Michelson began interviewing candidates to find a research staff member to do new research in biomedical engineering. I remember Swamy coming to see Dr. Michelson for the interview. He was dressed in a brown suit with a tie and beige shirt. His hair was worn back with light streaks of grey, he had a peppered mustache, and greeted me as we passed with a huge smile. He appeared very eager, with a mission and purpose.

Swamy, like my mom and I, was one of the few Indians who worked at UMDMJ. Since the chance meeting in Dr. Michelson's laboratory, I saw him again at various times passing in the hallway. Again the greetings were silent with simple exchanges of smiles. One day when I went home for dinner, I saw Swami seated at our dinner table! My mom, always generous and kind, had invited Swamy over to our house for dinner as she had bumped into him in the hallway. After dinner, Swamy and I sat down and spoke for the first time. He seemed to be on a mission, in a gentle way, to convince me to pursue science, and use my skills in science, mathematics and computer science to help the world.

In particular that conversation opened me to the world of pattern analysis and pattern recognition. I remember that conversation well. Swamy began by telling me about some research he and his sister had done in India. He used this research example to give me an example of how pattern analysis could be used. The example was rather unconventional compared to his conventional research assignments. He and his sister were very curious about seeing if there was a correlation or pattern to peoples palm prints and the onset of disease. In short, Swamy was interested in exploring if there was any truth in palmistry. He explained in detail how they had collected nearly 1000 palm prints from various people. They then created a database with each individual's name and their palm print. The palm print was a hand drawn sketch. They then had asked each person for their health history including any major diseases they had had or were suffering from. This health history was attached to the palm print.

Each palm print and the associated health history were reviewed manually to see if there were correlations between palm print features to health history. According to Swamy they had found some clear patterns. Certain palm print features had a high correlation to certain forms of disease. He provided me various examples of what he and his sister had discovered. Here was mathematics and modern science being applied to understand an ancient practice. Palmistry itself was a method of visual pattern recognition. Swamy's example intrigued me.

Swamy then provided me another example of how pattern recognition could be used. One of Swamy's passions was to model the human heart's electrophysiological behavior. He shared with me diagrams written on notepads at our dinning room table of the then current theories of how heart signals propagate. It seemed quite complex. He advised me to explore this field since there were many unknowns which would require someone good at both mathematics and computer science to model such phenomenon. Remember, this was at a time when the use of computers was just in their nascent state in the biological sciences.

That dinner conversation really got me thinking. While I had learned a lot of mathematics, was a star student in my high school, and was challenging myself in building the E-Mail System, something in my heart awoke. Swamy's conversation inspired me to explore how the skills I was developing could be applied to pattern analysis. Moreover, applying computing to biology seemed fascinating. Over the next several months, Swamy and I kept in touch and would have lunch together in the UMDMJ cafeteria. He was always encouraging of my work and kept asking me to work with him. In the Summer of 1981, I was close to completing a version of the first E-Mail System and would have more free time. I had just been accepted to M.I.T., and I promised Swamy to do a project with him, even if it was long distance from M.I.T.

During that first semester at M.I.T., the Institute had recreantly implemented UROP, Undergraduate Research Opportunity Program. One of the research opportunities was to work in pattern analysis for Tadoma. Tadoma is a method by which deaf-blind people communicate. Few understood how deaf-blind people were able to "listen" to someone else through the tactile method of putting their hand on someone's face. I mentioned this project to Swamy in a phone call and he encouraged me to pursue it since it would provide me some hands-on learning on pattern analysis project.

I kept Swamy posted on my activities at M.I.T. In one of our conversations, Swamy told me about another interesting project that involved heart electrophysiology relative to young infants. Apparently there was a phenomenon where young infants were dying in there sleep. This was called Sudden Infant Death Syndrome (SIDS). In SIDS, babies died in their sleep from what was known as an apnea. Today many hospitals and sleep labs test people for sleep apnea. At that time in the early 1980s, research in SIDS was just a new field. Swamy, through his collaborations, had acquired access to the best infant sleep data through Montefiore Hospital in New York City. This time series sleep data provided sleep states of thousands of babies as well as points in time at which they had the occurrence of an apnea. Swamy's thesis was that babies sleep states, or patterns of sleep states, may be indicative of the onset of an apnea.

Swamy gently prodded me to get involved in this project to build algorithms to test his hypothesis. Over the next several months, he provided me various books on Haar and Walsh Transforms as well as books on signals processing. Signals processing was a passion of Swamy's. He wanted me to learn as much as possible in this field for he knew that it would be a strong foundation for future work in pattern analysis. I as then a 17 year old had no idea of its long term importance. The Tadoma project I was working on at that time provided me experience in data acquisition but not in actually developing algorithms.

Swamy, to support my learning, advised me to write up sample programs using Haar and Walsh Transforms. I took his advice and learned a great deal about these two signals processing methodologies. Next year, Swamy provided me raw data of SIDS from Montefiore Hospital that he had acquired in 1977. He also provided me various papers he had written dealing with SIDS in his previous work. The earlier programming project I had done on Haar and Walsh Transforms was valuable to the SIDS project. In this project, my task was to review the data and see if I could find patterns of sleep states leading to the onset of an apnea.

The SIDS project was like solving a puzzle. Swamy taught me that babies have six states of sleep, whereas adults only have five. The data resembled up and down steps, each step being a sleep state. There were certain points which were marked when an apnea took place. The goal was to look backward from the point of the apnea to see if certain patterns sleep state correlated to when an apnea occurred.

I used the earlier Haar and Walsh Transforms to build pattern analysis methods to predict the onset of an apnea from the sleep waiting times.

While traveling back and forth between M.I.T. and UMDNJ, Swamy never micro managed me but was always there with patience to answer any of my questions. I worked on this for over a year. Some valuable results came from the analysis. The next year, the Fall of 1983, we had heard about the international conference to be held in Espoo, Finland. This conference on medical and biological engineering was the worlds largest such event. In the world of academia were as someone once said, "people fight over nothing", Swamy was unbelievably generous, again in retrospect. He wanted to include my results in a paper on SIDS, and where many in academia would have never thought about giving an undergraduate authorship on a paper, Swamy made me a co-author. In addition he invited me to come to Espoo to co-present the paper with him.

Going to Finland would be my first trip out of the United States to a foreign country besides India. Our paper was accepted for the Summer 1984 Conference. I was on my way to Finland. Attending that medical conference was an amazing experience. Twenty-seven years later today, I still remember that trip vividly. I was the youngest registering at the conference, but Swamy introduced me as his colleague from M.I.T. to all of the other scientists. For a 19-year old to go on a flight, attend banquets, hear scientific talks, travel the Finnish countryside, and hear a new language, was out of this world. Swamy had made that experience possible for me. He exposed me to a larger world, beyond science.

It was clear to me that scientists went to these conferences not just for the science but to experience other lands, local cuisines, and meet other people. Had Swamy not taken me on this trip, I would never have been exposed to this aspect of science. After coming back from that Conference, my enthusiasm for science and pursuing research in pattern analysis and pattern recognition skyrocketed.

During 1985–1994, to the beginning of my Ph.D. at M.I.T., I participated in numerous pattern recognition research projects. Swamy and I would talk from time to time on the phone, and I would keep him abreast on my work at MIT. He was always so very supportive, encouraging and uplifting. He was always positive and always ready to help. Swamy shared with me only a bit of his personal history. All I knew is that he had been in Holland for sometime, had gotten married to a Dutch woman, and had children. His moral support of my research activities helped me in my research pursuits. Among the projects I was involved in included: handwriting recognition, ultrasonic wave analysis, document analysis, image flow visualization, and a number of other pattern analysis projects.

From these various pattern analysis projects, there appeared to be a common set of strategies and methods that could be abstracted across all fields. That thought led to my Ph.D. thesis at M.I.T. entitled *Information Cybernetics*. In 1993, while in the middle of my Ph.D. work, I was invited to participate in a very interesting pattern analysis competition. This competition was not scientific in origin; it was from the U.S. Government. In particular, the Executive Office of the White House, with then President Bill Clinton, was looking for intelligent ways using computers to sort their E-Mail. At that time, the White House was receiving nearly

5,000 E-Mail per day, and this was before the introduction of the World Wide Web. There were 147 different categories. Student interns at the White House were manually reading each E-Mail and assigning them into one of the 147 categories. Categories included drugs, education, death threats, etc. The White House was interested in automatic filtering and sorting of E-Mail for two reasons.

I approached the White House competition with little knowledge natural language processing. My approach was engineering, where I used a hybrid method of employing nearly 19 different methods spanning feature extraction, clustering, to supervised and unsupervised learning. I was the only graduate student involved. The other five competitors were private and publicly traded companies. I won the competition.

I took time off in M.I.T. in 1994 to start EchoMail, a company for pattern analysis of electronic mail. We grew the company to nearly 300 employees worldwide by year 2000. I remember hearing from Swamy once during that time asking me for a letter of recommendation as he was taking a new job in the Midwest. It was weird for me to write a letter of recommendation for him. He had written several for me many decades before. Following that interaction I had not heard from him. It was my mom who a few years later informed me of his passing.

Hearing of Swamy's passing was sad. He had always been there and now was no more. I wondered who was with him, what his life had been during those past 10 years when I had lost touch. I recall Swamy and I sharing at our first dinner conversation thoughts about ancient Indian science, its mystical and scientific aspects, including the concept of Soul. There was an unstated agreement that Soul never dies. As I heard about Swamy's passing, I flashed back to that dinner conversation and knew his spirit would and had never died.

As I look back from the time of my meeting Swamy till today, 2010, my entire history of work, particularly my success with EchoMail, development of CytoSolve, and my recent research at M.I.T. to discover patterns of coherence that bridge traditional systems of medicine with modern systems biology, can all be traced to that dinner conversation with Swamy. As I write this last sentence, I am deeply moved and wish Swamy was here so we could have dinner again and I could thank him for the wondrous gift he gave a 16-year old nearly 30 years ago.

References

1. Hood L, Heath JR, Phelps ME, Lin B. Systems biology and new technologies enable predictive and preventative medicine. Science. 2004;306:640–3.
2. Ideker T, Lauffenburger D. Building with a scaffold: emerging strategies for high- to low-level cellular modeling. Trends Biotechnol. 2003;21:255–62.
3. Kitano H. Computational systems biology. Nature. 2002;420:206–10.
4. Palsson BO, Price ND, Papin JA. Development of network-based pathway definitions: the need to analyze real metabolic networks. Trends in Biotechnology. 2003;21:195–8.
5. Tomita M, Hashimoto K, Takahashi K, Shimizu TS, Matsuzaki Y, Miyoshi F, Saito K, Tanida S, Yugi K, Venter JC, Hutchison CA 3rd. E-CELL: software environment for whole-cell simulation. Bioinformatics. 1999;15:72–84.

6. Pennisi E. A low number wins the GeneSweep pool. Science. 2003;300:1484.
7. Hodgkin J. What does a worm want with 20, 000 genes? Genome Biology. 2001;2:1–4.
8. Putnam NH, Srivastava M, Hellsten U, Dirks B, Chapman J, Salamov A, Rokshar DS. Sea anemone genome reveals the gene repertoire and genomic organization of the Eumetazoan ancestor. Berkeley: Lawrence Berkeley National Laboratory; 2007.
9. Peri S, Navarro JD, Amanchy R, Kristiansen TZ, Jonnalagadda C, Surendranath V, Niranjan V, Muthusamy B, Gandhi TKB, Gronborg M, Ibarrola N, Deshpande N, Shanker K, Shivashankar HN, Pandey A. Development of human protein reference database as an initial platform for approaching systems biology in humans. Genome Research. 2003;13:2363–71.
10. Laue Mv. Kritische Bemerkungen zu den Deutungen der Photoframme von Friedich und Knipping. Physikalische Zeitschrift. 1913;14:421–3.
11. Patwardhan B, Warude D, Pushpangadan P, Bhatt N. Ayurveda and traditional Chinese medicine: a comparative overview. Oxford Journals Medicine Evidence-based Complementary and Alternative Medicine. 2005;2:465–73.
12. Subbarayappa BV. Siddha medicine: an overview. Lancet. 1997;350:1841–4.
13. Cannon WB. The wisdom of the body. New York: Norton; 1933.
14. Wiener N. Cybernetics or control and communication in the animal machine. Cambridge: The MIT Press; 1948.
15. Watson JD, Crick FH. Molecular structure of nucleic acids: a structure of deoxyribose nucleic acid. Nature. 1953;171:737–8.
16. Kitano H. Perspectives on systems biology. New Generation Computing. 2000;18:199–216.
17. Kitano H. Foundations of systems biology. Cambridge: The MIT Press; 2001.
18. Cuellar AA, Lloyd CM, Nielsen PF, Bullivant DP, Nickerson DP, Hunter PJ. An overview of CellML 1.1, a biological model description language. SIMULATION. 2003;79:740–7.
19. Le Novere N, Bornstein B, Broicher A, Courtot M, Donizelli M, Dharuri H, Li L, Sauro H, Schilstra M, Shapiro B, Snoep JL, Hucka M. BioModels database: a free, centralized database of curated, published, quantitative kinetic models of biochemical and cellular systems. Nucleic Acids Res. 2006;34:D689–91.
20. Snoep JL, Bruggeman F, Olivier BG, Westerhoff HV. Towards building the silicon cell: a modular approach. Biosystems. 2006;83:207–16.
21. Raczynski S. Differential inclusions in system simulation. Transactions of the Society for Computer Simulation. 1996;13:47–54.
22. Bulatwicz TF. Support for model coupling: an interface-based approach. Eugene: Department of Computer and Information Science, University of Oregon; 2006. p. 216.
23. Krueger CW. Software reuse. ACM Computing Surveys (CSUR). 1992;24:131–83.
24. Rajlich V, Wilde N. The role of concepts in program comprehension. In: 2002 international workshop on program comprehension. Los Alamitos, CA: IEEE Computer Society Press; 2002. p. 271–8.
25. Robinson S, Nance RE, Paul RJ, Pidd M, Taylor SJE. Simulation model reuse: definitions. benefits and obstacles., Simulation Modelling Practice and Theory. 2004;12:479–94.
26. Gianchandani EP, Brautigan DL, Papin JA. Systems analyses characterize integrated functions of biochemical networks. Trends in Biochemical Sciences. 2006;31:284–91.
27. Palsson B. Two-dimensional annotation of genomes. Nat Biotechnol. 2004;22:1218–9.
28. Papin JA, Hunter T, Palsson BO, Subramaniam S. Reconstruction of cellular signalling networks and analysis of their properties. Nat Rev Mol Cell Biol. 2005;6:99–111.
29. Hunter P, Borg T. Integration from proteins to organs: the Human Physiome project. Nature Reviews Molecular Cell Biology. 2003;4:237–43.
30. Hood L, Perlmutter RM. The impact of systems approaches on biological problems in drug discovery. Nat Biotechnol. 2004;22:1215–7.
31. Aderem A. Systems biology: its practice and challenges. Cell. 2005;121:511–3.
32. Takahashi K, Kaizu K, Hu B, Tomita M. A multi-algorithm, multi-timescale method for cell simulation. Bioinformatics. 2004;20:538–46.
33. Cerami EG, Bader GD, Gross BE, Sander C. cPath: open source software for collecting, storing, and querying biological pathways. BMC Bioinformatics. 2006;7:497.

34. Liu ET. Systems biology, integrative biology, predictive biology. Cell. 2005;121:505–6.
35. Hwang D, Smith JJ, Leslie DM, Weston AD, Rust AG, Ramsey S, de Atauri P, Siegel AF, Bolouri H, Aitchison JD, Hood L. A data integration methodology for systems biology: experimental verification. Proc Natl Acad Sci USA. 2005;102:17302–7.
36. Endy D, Brent R. Modelling cellular behaviour. Nature. 2001;409:391–5.
37. Sauro HM, Hucka M, Finney A, Wellock C, Bolouri H, Doyle J, Kitano H. Next generation simulation tools: the systems biology Workbench and BioSPICE integration. Omics. 2003;7:355–72.
38. Lindon JC, Holmes E, Nicholson JK. Metabonomics techniques and applications to pharmaceutical research & development. Pharm Res. 2006;23:1075–88.
39. Pecou E. Splitting the dynamics of large biochemical interaction networks. J Theor Biol. 2005;232:375–84.
40. Pennisi E. How will big pictures emerge from a sea of biological data? Science. 2005;309:94.
41. Bader JS, Chant J. Systems biology. When proteomes collide, Science. 2006;311:187–8.
42. Arkin AP, Fletcher DA. Fast, cheap and somewhat in control. Genome Biol. 2006;7:114.
43. Vaidehi N, Goddard WA. Atom-level simulation and modeling of biomacromolecules. In: Bower JM, Bolouri H, editors. Computational modeling of genetic and biochemical networks. Cambridge: MIT Press; 2001. p. 161–3.
44. White J. Two protein interactions are intractable using molecular dynamics. In: Ayyadurai S, editor. Personal communication. Cambridge; 2007.
45. Stultz C. Intractability of using atom-by-atom molecular dynamics for modeling biological pathways. In: Ayyadurai S, editor. Personal communication. Cambridge; 2007.
46. Keller EF. A clash of two cultures. Nature. 2007;445:603.
47. Noble D. Systems biology and the heart. Biosystems. 2006;83:75–80.
48. Duarte NC, Becker SA, Jamshidi N, Thiele I, Mo ML, Vo TD, Srivas R, Palsson BO. Global reconstruction of the human metabolic network based on genomic and bibliomic data. Proc Natl Acad Sci USA. 2007;104:1777–82.
49. Bhalla US. Understanding complex signaling networks through models and metaphors. Prog Biophys Mol Biol. 2003;81:45–65.
50. Hornberg JJ, Bruggeman FJ, Westerhoff HV, Lankelma J. Cancer: a systems biology disease. Biosystems. 2006;83:81–90.
51. Klipp E, Liebermeister W. Mathematical modeling of intracellular signaling pathways. BMC Neurosci. 2006;7(Suppl 1):S10.
52. Asthagiri AR, Lauffenburger DA. Bioengineering models of cell signaling. Annu Rev Biomed Eng. 2000;2:31–53.
53. Ma'ayan A, Jenkins SL, Neves S, Hasseldine A, Grace E, Dubin-Thaler B, Eungdamrong NJ, Weng G, Ram PT, Rice JJ, Kershenbaum A, Stolovitzky GA, Blitzer RD, Iyengar R. Formation of regulatory patterns during signal propagation in a Mammalian cellular network. Science. 2005;309:1078–83.
54. Lauffenburger DA. Cell signaling pathways as control modules: complexity for simplicity? Proc Natl Acad Sci USA. 2000;97:5031–3.
55. Bassingthwaighte JB, Chizeck HJ, Atlas LE, Qian H. Multiscale modeling of cardiac cellular energetics. Ann N Y Acad Sci. 2005;1047:395–424.
56. Alon U. Biological networks: the tinkerer as an engineer. Science. 2003;301:1866–7.
57. Kitney R, Dollery C. Systems biology: a vision for engineering and medicine. In: Klipp E, Liebermeister W, editor. Engineering. 2007.
58. Oltvai ZN, Barabasi AL. Systems biology. Life's complexity pyramid, Science. 2002;298:763–4.
59. Seeman NC, Belcher AM. Emulating biology: building nanostructures from the bottom up. Proc Natl Acad Sci USA. 2002;99(Supplement 2):6451–5.
60. Morgan JJ, Surovtsev IV, Lindahl PA. A framework for whole-cell mathematical modeling. J Theor Biol. 2004;231:581–96.
61. Sauro HM, Kholodenko BN. Quantitative analysis of signaling networks. Prog Biophys Mol Biol. 2004;86:5–43.

62. Mishra J, Bhalla US. Simulations of inositol phosphate metabolism and its interaction with InsP(3)-mediated calcium release. Biophys J. 2002;83:1298–316.
63. Klipp E, Nordlander B, Kruger R, Gennemark P, Hohmann S. Integrative model of the response of yeast to osmotic shock. Nat Biotechnol. 2005;23:975–82.
64. Hunter P, Smith N, Fernandez J, Tawhai M. Integration from proteins to organs: the IUPS Physiome Project. Mech Ageing Dev. 2005;126:187–92.
65. Alvarez-Vasquez F, Sims KJ, Hannun YA, Voit EO. Integration of kinetic information on yeast sphingolipid metabolism in dynamical pathway models. J Theor Biol. 2004;226:265–91.
66. Oda K, Kimura T, Matsuoka Y, Funahashi A, Muramatsu M, Kitano H. Map of the TLR signaling network. AfCS Research Reports. 2004;2:1–12.
67. Kitano H, Funahashi A, Matsuoka Y, Oda K. Using process diagrams for the graphical representation of biological networks. Nat Biotechnol. 2005;23:961–6.
68. Schulz M, Uhlendorf J, Klipp E, Liebermeister W. SBMLmerge, a system for combining biochemical network models. Genome Inform. 2006;17:62–71.
69. Akarsu E, Fox F, Furmanski W, Haupt T. WebFlow-high-level programming environment and visual authoring toolkit for high performance distributed computing. In: Proceedings of supercomputing '98: high performance networking and computing. IEEE Computer Society; 1998. p. 1–7.
70. Whelan G, Castleton KJ, Buck JW, Hoopes BL, Pelton MA, Strenge DL, Gelston GM, Kickert RN. Concepts of a framework for risk analysis in multimedia environmental systems (FRAMES). In: Laboratory PNN, editor. PNNL-11748. Richland: Pacific Northwest National Laboratory; 1997.
71. Neteler M, Mitasova H. Open source GIS: A GRASS GIS approach. Springer: Boston; 2004.
72. Dewey CF. In: Ayyadurai S, editor. Personal communication; 2006.
73. Quackenbush J, Stoeckert C, Ball C, Brazma A, Gentleman R, Huber W, Irizarry R, Salit M, Sherlock G, Spellman P, Winegarden N. Top-down standards will not serve systems biology. Nature. 2006;440:24.
74. Le Novere N, Finney A, Hucka M, Bhalla US, Campagne F, Collado-Vides J, Crampin EJ, Halstead M, Klipp E, Mendes P, Nielsen P, Sauro H, Shapiro B, Snoep JL, Spence HD, Wanner BL. Minimum information requested in the annotation of biochemical models (MIRIAM). Nat Biotechnol. 2005;23:1509–15.
75. Gilbert D, Fuss H, Gu X, Orton R, Robinson S, Vyshemirsky V, Kurth MJ, Downes CS, Dubitzky W. Computational methodologies for modelling, analysis and simulation of signalling networks. Brief Bioinform. 2006;7:339–53.
76. Brooks F. The mythical man month: essays in software engineering. Reading, MA: Addison Wesley; 1975.
77. Ayyadurai VAS, Dewey CF. CytoSolve: A scalable computational method for dynamic integration of multiple molecular pathway models. Cell Mol Bioeng. 2010; doi: 10.1007/s12195-010-0143-x.
78. Sato M, Taniguchi T, Tanaka N. The interferon system and interferon regulatory factor transcription factors—studies from gene knockout mice. Cytokine & Growth Factor Reviews. 2001;12:133–42.
79. Bocharaov. Mathematical model of antiviral immune response III. Influenza A virus infection. J Theor Biol. 1994;167:323–9.
80. Zi Z, Cho K, Sung M. In silico identification of the key components and steps in IFN-gamma induced JAK-STAT. FEBS. 2005;579:1101–8.
81. Hancioglu, B., Swigon, D., Clermont, G. A dynamical model of human immune response to influenza A virus infection. J Theor Biol. 2007;167:323–60.
82. Yamada S. Computer modeling of JAK/STAT signal transduction pathway. Genome Inform. 2001;12:282–3.
83. Cella M. Maturation, activation, and protection of dendritic cells induced by double-stranded RNA. J Exper Med. 1999;189:821–9.

84. Cooley M. Cytokine activity after human bone marrow transplantation: production of interferons by peripheral blood mononuclear cells from recipients of HLA-Identical sibling bone marrow transplants. J Immunol. 1987;138:3688–94.
85. Takauji. CpG-DNA-induced IFN- production involves p38 MAPKdependent STAT1 phosphorylation in human plasmacytoid dendritic cell precursors. J Luekocyte Biol. 2002;72:1011–19.
86. Kimmel AR, Parent CA. The signal to move: D. discoideum go orienteering. Science. 2003;300:1525–7.
87. Lauffenburger DA. Four M's of systems biology. Cambridge: MIT; 2003.